Securing an Enterprise

Maximizing Digital Experiences through Enhanced Security Measures

Anirudh Khanna

Apress®

Securing an Enterprise: Maximizing Digital Experiences through Enhanced Security Measures

Anirudh Khanna
Primus Global Services
Plano, TX, USA

ISBN-13 (pbk): 979-8-8688-1028-2   ISBN-13 (electronic): 979-8-8688-1029-9
https://doi.org/10.1007/979-8-8688-1029-9

Copyright © 2024 by Saurav Bhattacharya

This work is subject to copyright. All rights are reserved by the Publisher, whether the whole or part of the material is concerned, specifically the rights of translation, reprinting, reuse of illustrations, recitation, broadcasting, reproduction on microfilms or in any other physical way, and transmission or information storage and retrieval, electronic adaptation, computer software, or by similar or dissimilar methodology now known or hereafter developed.

Trademarked names, logos, and images may appear in this book. Rather than use a trademark symbol with every occurrence of a trademarked name, logo, or image we use the names, logos, and images only in an editorial fashion and to the benefit of the trademark owner, with no intention of infringement of the trademark.

The use in this publication of trade names, trademarks, service marks, and similar terms, even if they are not identified as such, is not to be taken as an expression of opinion as to whether or not they are subject to proprietary rights.

While the advice and information in this book are believed to be true and accurate at the date of publication, neither the authors nor the editors nor the publisher can accept any legal responsibility for any errors or omissions that may be made. The publisher makes no warranty, express or implied, with respect to the material contained herein.

    Managing Director, Apress Media LLC: Welmoed Spahr
    Acquisitions Editor: Aditee Mirashi
    Development Editor: James Markham
    Coordinating Editor: Kripa Joseph

Cover designed by eStudioCalamar

Cover image by Freepik (www.freepik.com)

Distributed to the book trade worldwide by Apress Media, LLC, 1 New York Plaza, New York, NY 10004, U.S.A. Phone 1-800-SPRINGER, fax (201) 348-4505, e-mail orders-ny@springer-sbm.com, or visit www.springeronline.com. Apress Media, LLC is a California LLC and the sole member (owner) is Springer Science + Business Media Finance Inc (SSBM Finance Inc). SSBM Finance Inc is a **Delaware** corporation.

For information on translations, please e-mail booktranslations@springernature.com; for reprint, paperback, or audio rights, please e-mail bookpermissions@springernature.com.

Apress titles may be purchased in bulk for academic, corporate, or promotional use. eBook versions and licenses are also available for most titles. For more information, reference our Print and eBook Bulk Sales web page at http://www.apress.com/bulk-sales.

Any source code or other supplementary material referenced by the author in this book is available to readers on GitHub (https://github.com/Apress). For more detailed information, please visit https://www.apress.com/gp/services/source-code.

If disposing of this product, please recycle the paper

Table of Contents

About the Author xvii
Contributing Authors xix
About the Technical Reviewer xxiii

Part I: Introduction to Cybersecurity 1

Chapter 1: Introduction to Cybersecurity 3
Define Security in the Context of Digital Information 3
    Confidentiality 4
    Integrity 4
    Availability 5
    Why These Principles Are Essential 5
The Necessity of Protecting Data 11
Conclusion 17
References 18

Chapter 2: Threat Landscape 19
Overview of Current and Emerging Threats 19
    Current Threats 20
    Emerging Threats 28
    Mitigation Strategies 31
Cybercrime, Malware, and State-Sponsored Attacks 33
    Cybercrime 33
    Malware Campaigns 34

TABLE OF CONTENTS

    State-Sponsored Cyber Warfare 36

        Distinguishing Between Cybercrime and StateSsponsored Cyber Warfare... 40

    Summary.. 42

    References 43

Chapter 3: Security Principles 47

    Detailed Exploration of Encryption 47

        Types of Encryptions 47

    Authentication and Access Controls 54

        Authentication 54

        Access Control Models 58

        Implementation of Authentication and Access Control 62

    Integrity Checks and Other Security Measures 64

    Summary.. 70

    References 71

Chapter 4: Authentication in Cybersecurity 75

    Introduction 75

        Definition of Authentication 76

        Importance of Authentication in Cybersecurity 77

    Types of Authentication 78

        Password-Based Authentication 79

        Two-Factor Authentication 81

        Biometric Authentication 82

        Multifactor Authentication 83

    Common Authentication Vulnerabilities 84

        Password Weaknesses 84

        Credential Theft 85

TABLE OF CONTENTS

Social Engineering Attacks 86
Brute Force Attacks 86
Best Practices for Secure Authentication..87
Strong Password Policies 88
Regular Password Updates 89
User Education and Awareness 90
Implementing Multifactor Authentication 91
Conclusion 92
References..94

Part II: Ransomware Basics and Prevention 95

Chapter 5: Introduction to Ransomware..97
What Is Ransomware? 97
Evolution of Ransomware 97
Impact of Ransomware on Individuals and Organizations 99
Timeline of Prominent Ransomware Attacks 100
References..103

Chapter 6: Ransomware Lifecycle 105
Phase 1: Reconnaissance and Target Selection.....................................105
Phase 2: Initial Access 107
Spear Phishing 108
Unpatched Software 108
Phase 3: Lateral Movement and Privilege Escalation 110
Phase 4: Deployment of Ransomware Payload......................................112
Phase 5: Encryption and Impact 114
Phase 6: Extortion and Communication 116
Communication Phase: Navigating a Minefield 117

TABLE OF CONTENTS

Chapter 7: Ransomware Prevention 119
Regular Software Updates and Patch Management 119
Employee Education and Training 122
Implementing Robust Endpoint Security Solutions 124
Data Backup and Disaster Recovery Planning 126
Network Segmentation and Access Controls 129
Email and Web Security Measures 132
Developing a Comprehensive Prevention Strategy 135

Part III: Ransomware Detection and Response 139

Chapter 8: Early Detection Techniques 141
Unveiling the Challenge of Ransomware Detection 141
Strategies for Enhanced Detection 142
Elevating Cybersecurity Literacy 143
Familiarizing with the Hallmarks of Ransomware 144
Empowering the Workforce 145
Implementation and Benefits 145
Challenges in Process Anomaly Detection 150
Implementation and Techniques 150
Benefits of Mastering Process Anomaly Detection 151
The Role of Monitoring 153
Importance of Documenting Legitimate Behaviors 153
Investigating Anomalies 154
Challenges and Solutions 154
The Significance of Vigilance 156
Investigative Techniques 157
References 160

TABLE OF CONTENTS

Chapter 9: Incident Response 163
    The Core of Threat Detection 164
    Leveraging Tools and Technologies..164
    Analytical Exploration 170
    Incident Response Framework 170
    Real-World Applications 172
    Conclusion 172
    References...173

Chapter 10: Threat Intelligence 175
    Introduction...175
        Intelligence...176
        Cyberthreat..177
        Cyberthreat Intelligence 178
        Evolution of Threat Intelligence 178
        Why Threat Intel Is Useful 180
    References...191

Part IV: Ransomware Recovery Strategies 197

Chapter 11: Backup and Restore 199
    Importance of Regular Backups...199
        Major Backup Methods..203
    Secure Backup Practices for Organizations................................206
        Emerging Practices in Backup and Restoration Within Organizations........222
    Summary...225
    References...226

TABLE OF CONTENTS

Chapter 12: Ransomware Recovery Framework 229
Introduction to Ransomware Recovery 229
    Identify the Attack 230
    Isolate Infected Systems 231
    Assess the Damage 232
    Report the Incident 233
    Identify the Ransomware Variant 234
    Decide on a Course of Action 234
    Restore from Backups 236
    Rebuild and Harden Systems 237
    Review and Improve Security 238
Identifying and Isolating Infected Systems 239
    Network Segmentation 239
    Endpoint Detection and Response (EDR) 240
    Intrusion Detection/Prevention (IDPS) 242
    Security Information and Event Management (SIEM) 244
    User Behavior Analytics (UBA) 246
    Honeypots 247
    Incident Response Playbooks 248
Summary 249
References 250

Chapter 13: Negotiating with Attackers 255
Overview 255
Key Statistics on Ransomware Attacks in the United States, Europe, and Globally 256
Legal and Ethical Considerations 260
    Legal Considerations 260
    Moral Considerations 262

Balancing Risks and Benefits 264
Factors to Consider Before Making Negotiations 266
    Data Criticality 267
    Ransom Payment Feasibility 267
    Data Recovery Alternatives 267
    Operational Disruption 268
    Third-Party Consequences 268
    Insurance Considerations 268
    Public Relations and Brand Equity 269
Seeking Legal Counsel 269
Working with Law Enforcement 271
    Benefits of Collaboration 271
    Logistics of Collaboration 272
    Building Trust and Communication 272
    Challenges in Collaboration 273
Summary 273
References 275

Chapter 14: Rebuilding and Strengthening Security Posture 283

Cyberattack-Learning from the Incidents 283
Implementing Security Enhancements 289
    Policy Adjustments 290
    Cybersecurity Framework 290
    Create a Cybersecurity-Centric Culture 292
    Technological Upgrades 293
    Monitoring and Managing Threats Real-Time 295
    Penetration Testing and Vulnerability Scanning 296
Summary 297
References 297

TABLE OF CONTENTS

Part V: Real-World Perspectives 301

Chapter 15: Case Studies in the E-commerce Industry 307
Overview of Ransomware Attacks in the E-commerce Industry 307
Selected Case Studies of Ransomware in the E-commerce Industry 309
Case Study One: Tupperware.com Ransomware Attack in 2021 309
Introduction to the Case 309
Detailed Analysis 310
Responses and Challenges to their Implementation 311
Effectiveness of the Strategies 311
Case Study Two: Ransomware Attack on Online Vendor X-Cart in October 2020 312
Introduction to the Case 312
Detailed Analysis 313
Responses and Challenges to Their Implementation 314
Effectiveness of the Strategies 315
Case Study Three: VF Corporation's Ransomware Attack on December 13, 2023 315
Introduction to the Case 315
Detailed Analysis 316
Response and Challenges to Their Implementation 317
Effectiveness of the Strategies 318
Case Study Four: Staples Ransomware Attack in December 2023 318
Introduction to the Case 318
Detailed Analysis 319
Response and Challenges Implementing Strategies 320
Effectiveness of Strategies 321
References 321

TABLE OF CONTENTS

Chapter 16: Ransomware, Inc.: The Business and Economics of Digital Extortion 329

Introduction: The Ransomware Economy 329

Inside the Ransomware Marketplace 330

From Petty Crime to Big Business: The Evolution of Ransomware 330
The Ransomware Value Chain: Players, Roles, and Profit Shares 331
Tailoring the Ransom Note: The Art and Science of Pricing Extortion 333
Attack Methodology: Infiltration, Exfiltration, Encryption, and Destruction 334
The Emergence of Ransomware as a Service 336
Ransomware Revenue and Profitability 336

The Minds Behind the Malware: Incentives of Ransomware Actors 337

The Profit Motives and Risk-Return Tradeoffs of Ransomware Actors 337
The Role of Cryptocurrency in the Ransomware Ecosystem 338
Deterring Ransomware: The Limits of Economic Incentives 339

Ransomware and the Bottom Line: The Business Costs of Digital Extortion.....340

The Direct Financial Costs of Ransomware 340
The Indirect Costs and Long-Term Impacts of Ransomware 341
The Business Case for Investing in Ransomware Prevention and Mitigation 342

Hacking the Hackers: Economic Strategies to Combat Ransomware 343

Reducing the Profitability of Ransomware 344
Increasing the Costs and Risks for Ransomware Actors 345
Building Resilience and Reducing the Impact of Ransomware 346
Case Study: The Colonial Pipeline Attack and the US Government's Response 348

Weathering the Storm: Lessons in Organizational Resilience 352

Developing a Comprehensive Incident Response Plan 352
Building a Culture of Security Awareness and Responsibility 352

TABLE OF CONTENTS

    Embracing Continuous Learning and Improvement 353
    Conclusion: Navigating the Future of Ransomware 354
    References 356

Chapter 17: Case Studies in Confidential Computing 361
    Overview of Confidential Computing 361
    Real-World Use Cases and Scenarios 362
    Challenges and Solutions in Implementation 363
    Impact on Data Privacy and Security 364
    Lessons Learned and Future Directions 365

Chapter 18: Case Studies in Cloud Computing 367
    Evolution of Cloud Computing and Its Security Challenges 367
    Noteworthy Cloud Security Case Studies 369
    Risk Mitigation Strategies and Best Practices 370
    Impact of Cloud Adoption on Organizations 372
    Emerging Trends and the Road Ahead 373
    Conclusion 375
    Multiple Choice Questions 376
        Answers 382

Chapter 19: Case Studies in Enterprise Security Architecture 385
    Understanding Enterprise Security Architecture 386
    Designing and Implementing Secure Enterprise Solutions 389
    Case Studies Showcasing Successful Security Architectures 392
    Case Study 1: Banking Sector 392
        Multiple Choice Questions 395
        Answers 401

TABLE OF CONTENTS

Chapter 20: Case Studies in Energy 403

Overview of Ransomware Attacks in the Energy Industry 403

Selected Case Studies of Ransomware Attacks in the Energy Industry 405

    Case Study 1: Nordex Group SE and Deutsche Windtechnik Ransomware Attacks 405

        Introduction to the Case 405

        Detailed Analysis 407

        Responses and Challenges to Their Implementation 407

        Effectiveness of the Strategies 408

    Case Study 2: Saudi Aramco's Petro Rabigh Complex Ransomware Attack 409

        Introduction to the Case 409

        Detailed Analysis 410

        Response and Challenges to Their Implementation 410

        Effectiveness of Strategies and Lessons Learned 411

    Case Study 3: Colonial Oil Pipeline Attack in the United States 412

        Introduction to the Case 412

        Detailed Analysis 413

        Response and Challenges to Their Implementation 414

        Effectiveness of the Strategies Used 415

    Case Study 4: Ransomware Attack on the Amsterdam–Rotterdam–Antwerp Refining Hubs 416

        Introduction to the Case 416

        Detailed Analysis 417

        Response and Challenges to Their Implementation 417

        Effectiveness of Ransomware Recovery Strategies 418

References 418

TABLE OF CONTENTS

Chapter 21: Securing Digital Foundations: A Point of View on Cybersecurity in Healthcare **429**

Key Components of Cybersecurity in Healthcare 430

Threat Landscape in Healthcare 432

Security Principles in Healthcare 434

Ransomware in Digital Healthcare Landscape 435

    Understanding the Entry Points of Ransomware 436

    Impact on Healthcare: The Consequences of Ransomware in Digital Healthcare 437

Ransomware Prevention in a Digital Healthcare System 439

Ransomware Detection and Response: Early Detection Techniques in Healthcare Administration and Management 443

    Analyzing the Data Trends in Healthcare Ransomware Attacks 451

Incident Response: The Lifeline of Digital Healthcare 453

Securing the Digital Pulse: Advanced Threat Intelligence Strategies for Healthcare 456

Future of Digital Security in Healthcare 461

    The Digital Transformation Landscape in Healthcare 461

    Future Trends in Healthcare Cybersecurity 463

    Transformative Security Technologies 465

    Case Examples and Empirical Evidence 470

Conclusion 474

References 475

Part VI: Future Trends in Cybersecurity **479**

Chapter 22: Future Trends in Digital Security **481**

Future Trends and Anticipations in Digital Security 481

    Artificial Intelligence and Machine Learning 481

    Zero Trust Architecture 482

TABLE OF CONTENTS

- Edge Computing Strategies 483
- Cloud Computing Strategies 484
- Quantum Computing in Data Encryption 485
- The Role of Encryption in Futuristic Digital Security 485
  - Post-quantum Encryption 485
  - Homomorphic Encryption 486
  - Encrypted Machine Learning 487
- Biometric Authentication and Identity Management 488
  - AI-Enabled Multimodal Biometrics 488
  - Human–Machine Adaptation 489
- Blockchain Technology and Security Applications 490
- Cybersecurity in the Internet of Things (IoT) 491
- Security in a Remote Work Environment 492
- Shortage of Skilled Talent in Digital Security 493
- The Digital Transformation Landscape in Healthcare 495
  - Consumer-Centric Approach 495
  - Interim Milestones and Value Measurement 496
  - Talent and Data Challenges 496
- Summary 496
- References 497

Index **505**

About the Author

Anirudh Khanna is a distinguished technical and thought leader in the fields of backup and recovery, disaster recovery, and ransomware attack recovery. With over 16 years of experience, Anirudh has successfully led initiatives to safeguard data and ensure business continuity for several Fortune 500 companies.

Anirudh is widely recognized for his expertise in ransomware protection and recovery. Notably, he led the implementation of a ransomware recovery solution for one of the largest gas and electricity companies in the United States, ensuring compliance with the TSA directive. His efforts significantly improved the company's resilience against cyberattacks, reduced recovery times, and strengthened its overall cybersecurity posture.

In addition to his technical achievements, Anirudh serves as the Senior Vice President of the New World Foundation, where he led the development of a patented solution for conducting cybersecurity assessments tailored to small businesses. This groundbreaking innovation integrates modules for cyberattack prevention, recovery, privacy reviews, risk assessment, and security audits, addressing the unique challenges faced by smaller enterprises in today's threat landscape.

ABOUT THE AUTHOR

As a prolific author, Anirudh has published over 20 research papers in reputed journals and presented at more than seven international conferences. His contributions to the advancement of ransomware defense mechanisms have garnered numerous prestigious awards, including the Stevie, Globee, and Titan Awards for technological excellence.

For the past seven years, Anirudh has played a pivotal role in providing business continuity services for critical infrastructure at one of the largest utility companies in the United States. He is also a Senior Member of IEEE, regularly reviews research papers for highly reputed journals and conferences and is a sought-after expert in data protection and cyber recovery.

Anirudh's extensive body of work, including articles and research papers, highlights his capabilities as a visionary in data protection and cyber recovery, cementing his status as a respected authority in the industry.

Contributing Authors

Saurav Bhattacharya is a distinguished researcher and author with extensive expertise in account registration systems, digital identity, and cybersecurity. With a background in computer science from IIT Kharagpur and a career at Microsoft, Saurav has been instrumental in advancing technology solutions that address global challenges. As the founder of an online security firm, SuperChargePlus, and president of the New World Foundation, he brings a wealth of knowledge and leadership to the peer-reviewed journal IJGIS.

With a master's degree in Computer Science from Indiana University Bloomington and multiple cloud certifications, **Ayisha Tabbassum** is an Onsite Lead for Cloud Operations and Multi-Cloud Architecture at Otis Elevator Co. She designs, automates, provisions, and secures Azure, AWS, and GCP infrastructure for various business domains and customer needs. She is also the founder and CEO of One Stop for Cloud, an Edtech company with the motto of providing simplified learning solutions for five major cloud platforms such as AWS, Azure, GCP, OCI, and IBM. She is a conference speaker on AI and cloud technologies. She has extensive work experience in using the most sophisticated cloud platforms, such as AWS, Azure, GCP, and IBM, to create scalable, reliable, and cost-effective solutions. She is also responsible for reporting and addressing the security vulnerabilities in the Azure security center, Wiz, and AWS Security Hub and designing and implementing policy add-ons to enhance security. In addition to her cloud engineering and architecture skills, she has a strong background in infrastructure automation and CI-CD application deployments, using technologies such as Git, Gitlab, Jenkins, Ansible, Docker, Kubernetes, Openshift, Dynatrace, Splunk, Prometheus, Grafana, Sitescope, Nagios, ELK, and Azure Monitor. She has applied these skills in diverse domains, such as e-commerce, retail, big data, and security, delivering high-quality solutions that meet business requirements and customer expectations. She is passionate about learning new technologies and staying updated with the latest trends and best practices in cloud computing and DevOps. She is also motivated by collaborating with cross-functional teams and stakeholders and contributing to the organization's goals and vision.

Fardin Quazi is a renowned expert in digital and business transformation within the healthcare domain, with 19+ years of extensive global experience in healthcare technology, management and admin solutions, robotics and intelligent process automation, AI-ML, and digital technology-based business transformation solutions. Fardin is working as an associate director—Business Solutions, with Cognizant Technology Solutions, US Corp. He is a Certified Professional of the Academy for Healthcare Management, issued by American Health Insurance Plans. He holds an MBA in Information Systems and Operations and a bachelor's in Electrical Engineering. Fardin is volunteering as the Senior Vice President of Ethics Standards and Compliance at the New World Foundation and serves as an Editorial Board Member for the *International Journal of Global Innovations and Solutions*. He is currently living in Dallas, TX, with his family.

CONTRIBUTING AUTHORS

Harshavardhan Nerella is a distinguished cloud engineer with over seven years of experience, complemented by two master's degrees from prestigious universities in the United States. He has a robust background in cloud computing, Cloud Native Solutions and Kubernetes. He is deeply involved in research and technical community contributions. He has published research papers in esteemed journals and conferences and authored articles featured in DZone's Spotlight section. His commitment to the field extends to his roles as a reviewer for various conferences and journals, and he has acted as a judge for prestigious competitions such as Princeton Research Day and Technovation. Recognized as a Top Cloud Computing Voice on LinkedIn, he is also a highly sought-after mentor and interview preparation guide on ADPList, where he is ranked in the top 1% of mentors.

Varun Garde is a business strategy and analytics leader specializing in pricing and revenue management. He is currently Director of Monetization and Business Planning at Microsoft. He holds an MBA from Ross School of Business, University of Michigan—Ann Arbor, and has served in pricing and revenue leadership across B2C and B2B industries.

xxi

CONTRIBUTING AUTHORS

Dhruv Kumar Seth is a strategic e-commerce solutions architect and performance optimization specialist with over 16 years of experience in designing, developing, and deploying scalable, high-performance systems for the retail industry. His expertise spans microservices architecture, distributed systems, API integration, caching solutions, and performance tuning, enabling businesses to achieve enhanced scalability, operational efficiency, and customer satisfaction. Dhruv's contributions with industry leaders like Walmart and Apple Inc. have driven impactful innovations, including multi-tenant SaaS microservices applications and next-generation e-commerce platforms. A passionate mentor and thought leader, he actively shares his knowledge through technical councils, industry conferences, and articles, including a featured piece on DZone. Dhruv's forward-looking vision emphasizes technological advancements, mentoring the next generation of engineers and pioneering innovative solutions to drive the retail industry forward.

About the Technical Reviewer

Sagar Sidana is a principal software engineer at McKinsey & Company, based in Dallas, TX. With extensive experience in leading software development teams and implementing complex software solutions, Sagar has a proven track record of driving digital transformation initiatives across various industries. He excels in providing technical leadership throughout the software development lifecycle, ensuring the delivery of high-quality software products.

At McKinsey & Company, Sagar has designed and implemented a cutting-edge SaaS solution that empowers businesses to measure, analyze, and reduce their carbon footprint. By leveraging advanced data analytics and machine learning algorithms, his solutions integrate seamlessly with existing company systems, enabling data-driven decision-making for environmental sustainability.

Previously, Sagar served as a product architect at Deloitte Consulting LLP, where he designed a SaaS solution for an insurance marketplace, streamlining the process of finding and purchasing insurance for state citizens. His expertise includes utilizing cutting-edge technologies to create scalable and robust architectures.

Sagar's career also includes roles as an enterprise applications architect at MAPFRE Insurance, a senior consultant in digital and emerging technologies at Ernst & Young (EY), and a technology lead at

ABOUT THE TECHNICAL REVIEWER

Infosys. Throughout these positions, he has demonstrated a strong ability to define technology visions, conduct feasibility analyses, and lead teams to achieve business goals.

Sagar holds a Bachelor of Engineering degree from Maharshi Dayanand University and is a certified AWS Solutions Architect and ScrumMaster. His technical skills span platform architecture, software engineering, cloud infrastructure, and data platforms & solutions.

In addition to his technical prowess, Sagar is recognized for his leadership and communication skills. He has received several honors and awards, including the Applause Award from Deloitte, the Breakaway Award from MAPFRE, Culture Coin Winner from EY, and PRIMA award from Infosys. Sagar is committed to fostering a culture of innovation, collaboration, and continuous learning within his teams, and he actively mentors and coaches software engineers.

Sagar is passionate about leveraging AI technologies for process optimization and predictive analytics, contributing to advancements in life sciences and environmental sustainability.

For more information, connect with Sagar on LinkedIn at `linkedin.com/in/sagarsidana`.

PART I

Introduction to Cybersecurity

Chapter 1: Introduction to Cybersecurity

- **Define Security in the Context of Digital Information:** Explains the fundamental concepts of cybersecurity, including confidentiality, integrity, and availability. Discusses why these principles are essential for protecting digital information.
- **The Necessity of Protecting Data:** Delves into the reasons data protection is critical, including privacy concerns, financial implications, and the potential for identity theft.

Chapter 2: Threat Landscape

- **Overview of Current and Emerging Threats:** Provides a comprehensive overview of the types of threats that exist in the digital world, including viruses, worms, trojans, and advanced persistent threats (APTs).
- **Cybercrime, Malware, and State-Sponsored Attacks:** Discusses the motivations behind different types of attacks, distinguishing between cybercrime, malware campaigns, and state-sponsored cyber warfare.

PART I   INTRODUCTION TO CYBERSECURITY

Chapter 3: Security Principles

- **Detailed Exploration of Encryption:** Introduces the concept of encryption and its critical role in protecting data. Explains symmetric and asymmetric encryption, along with real-world applications.

- **Authentication and Access Controls:** Covers the basics of how authentication and access controls work to protect systems and data. Examines password policies, biometric authentication, and access control models.

- **Integrity Checks and Other Security Measures:** Discusses mechanisms like hashing and digital signatures used to ensure data integrity. Also touches on additional security measures like firewalls and antivirus software.

Chapter 4: Authentication in Cybersecurity

- **Understanding Authentication Mechanisms:** Explores the various authentication mechanisms, from traditional passwords to biometric and hardware-based authentication.

- **Password-Based, Biometric, and Multifactor Authentication:** Details the strengths and weaknesses of different authentication methods and the importance of multifactor authentication (MFA).

- **Challenges and Best Practices in Authentication:** Addresses common challenges in implementing strong authentication measures and offers best practices.

- **Future Trends in Authentication Technologies:** Looks at emerging trends and technologies in authentication, such as behavioral biometrics and decentralized identity systems.

CHAPTER 1

Introduction to Cybersecurity

Author:

Anirudh Khanna

In the digital era, our lives are intertwined with the internet and technology more than ever before. From personal communications to financial transactions and national infrastructure, everything is digitized. This digital revolution has brought unparalleled conveniences, but it has also introduced new vulnerabilities and threats. Cybersecurity, therefore, has become an indispensable aspect of our digital existence, safeguarding our information and systems from malicious attacks. This chapter lays the groundwork for understanding the fundamental concepts of cybersecurity, emphasizing the significance of data protection through the principles of confidentiality, integrity, and availability.

Define Security in the Context of Digital Information

The CIA Triad of confidentiality, integrity, and availability is foundational to understanding and implementing cybersecurity strategies. This model serves as a guiding framework for organizations to protect their digital information and assets effectively. Let's delve deeper into each of these principles and understand their significance in the realm of digital security.

CHAPTER 1   INTRODUCTION TO CYBERSECURITY

Confidentiality

Confidentiality is about ensuring that information is not disclosed to unauthorized individuals, entities, or processes. In the digital world, where data breaches are increasingly common, maintaining confidentiality is crucial. It's not just about keeping secrets; it's about protecting personal data, proprietary information, and sensitive communications that, if exposed, could lead to financial loss, reputational damage, or even legal consequences.

Encryption is a primary tool for maintaining confidentiality. By transforming readable data into an unreadable format, encryption ensures that even if data are intercepted, they remain inaccessible to unauthorized users. Access control measures, such as passwords, biometric verification, and digital certificates, further ensure that only those with the right credentials can access sensitive information.

Integrity

Integrity revolves around the assurance that information remains unchanged from its original state, barring any updates by authorized parties. This principle protects data from unauthorized modifications, deletions, or fabrication, ensuring that the information an organization relies upon is accurate and reliable. The consequences of compromised integrity can range from minor inconveniences to catastrophic business failures, depending on the nature of the altered data.

To safeguard integrity, organizations employ cryptographic hash functions, digital signatures, and version control mechanisms. These tools verify the authenticity of data and detect any unauthorized alterations, providing a secure means of ensuring that information remains trustworthy and untampered.

Availability

Availability ensures that information and resources are accessible to authorized users when needed. This principle addresses the resilience of systems against various forms of disruption, from cyberattacks like DDoS, which can overload services with traffic, to natural disasters that could cripple physical infrastructure.

To enhance availability, organizations implement redundant systems, data backups, and disaster recovery plans. These measures ensure that even in the face of attacks or failures, systems can recover quickly, minimizing downtime and maintaining continuous access to critical information and services.

Why These Principles Are Essential

The CIA Triad principles are essential because they address the comprehensive needs of digital information security. Confidentiality breaches can lead to significant legal, financial, and reputational damage. Integrity violations can undermine trust in data, leading to erroneous decisions and potential harm. Availability issues can disrupt operations, leading to financial losses and eroded customer trust.

Together, these principles guide organizations in developing a holistic cybersecurity strategy that protects against a wide range of threats. By prioritizing confidentiality, integrity, and availability, organizations can not only defend against cyberattacks but also foster trust among customers and stakeholders, ensuring the ongoing success and resilience of their operations in a digital world.

Scenario: A Healthcare Provider and EHR Management

To illustrate the CIA Triad principles in action, let's examine a hypothetical real-world scenario involving a healthcare provider that manages electronic health records (EHRs).

CHAPTER 1   INTRODUCTION TO CYBERSECURITY

Confidentiality in Action

The healthcare provider maintains detailed electronic health records for all its patients. These records contain sensitive information, including personal identification details, medical histories, treatment plans, and billing information. To uphold confidentiality, the provider employs encryption for data at rest and in transit. This means that even if a cybercriminal were to intercept the data, it would be unreadable and useless without the encryption keys. Additionally, access to these records is strictly controlled with secure authentication methods, ensuring that only authorized medical personnel can view or modify patient information.

Integrity in Action

To maintain the integrity of the EHRs, the healthcare provider uses digital signatures and checksums for all records. Every time a record is updated, a digital signature is applied, verifying the identity of the person making the change, and a checksum is calculated to ensure that the data have not been altered during transmission. This approach ensures that the records are accurate and have not been tampered with, whether by internal mistakes or external interference. It helps in preventing misdiagnosis or incorrect treatment that could result from altered patient data.

Availability in Action

The healthcare provider understands the critical nature of having EHRs available 24/7 for patient care. To ensure availability, the provider has implemented redundant systems and regular backups. In the event of a system failure or a cyberattack like a DDoS, these redundant systems and backups allow for a swift recovery, minimizing downtime. Furthermore, the provider has a comprehensive disaster recovery plan in place to quickly restore access to records in the event of a natural disaster, ensuring that medical personnel can continue to provide care without significant delays.

The Impact of Adhering to the CIA Triad

In this scenario, by adhering to the principles of the CIA Triad, the healthcare provider not only protects sensitive patient information but also ensures that the data remain accurate and accessible when needed. This commitment to cybersecurity fosters trust among patients, who feel confident that their private information is protected and that the medical care they receive is based on reliable data. Additionally, it shields the provider from potential legal and financial repercussions associated with data breaches, data tampering, and system downtime.

Advanced Applications of the CIA Triad
Beyond Basic Encryption: Zero Trust Architecture

In enhancing confidentiality, organizations are increasingly adopting the Zero Trust security model. This model operates on the principle that no entity inside or outside the network is trusted by default, and verification is required from everyone trying to access resources in the network. This approach minimizes the potential for unauthorized access and significantly enhances data confidentiality. It extends the concept of encryption by requiring continuous validation of credentials and implementing least-privilege access, ensuring that users and devices only have access to the information and resources necessary for their roles.

Ensuring Data Integrity Through Blockchain

The integrity aspect of the CIA Triad is seeing innovative applications through technologies like blockchain. Originally devised for digital currencies, blockchain provides a decentralized and immutable ledger, perfect for maintaining the integrity of transactional data or any data records. Its application in supply chain management, for example, ensures that every step — from production to delivery — is recorded and unalterable, preventing fraud and ensuring the authenticity of products.

Cloud Services and Availability

The shift toward cloud computing has significantly impacted the availability principle of the CIA Triad. Cloud providers offer highly resilient infrastructure and services, ensuring data and applications are available even in the event of major hardware failures or natural disasters. The scalability of cloud services means that resources can be adjusted based on demand, ensuring that spikes in usage do not result in system downtime. Moreover, cloud providers invest heavily in cybersecurity, further protecting against attacks that could threaten availability.

Challenges and Considerations

While the CIA Triad provides a robust framework for cybersecurity, its implementation is not without challenges. Balancing these principles can sometimes lead to trade-offs. For example, increasing data encryption (confidentiality) can impact system performance, potentially affecting availability. Similarly, stringent access controls enhance confidentiality but can complicate legitimate access, impacting the user experience.

Moreover, the dynamic nature of cyber threats requires that the approaches to protecting confidentiality, integrity, and availability evolve continuously. Organizations must stay informed about the latest cybersecurity trends and threats and be prepared to update their security practices regularly.

The Human Factor

An often overlooked but critical aspect of the CIA Triad is the role of people. Human error remains one of the largest vulnerabilities in cybersecurity. Educating employees about security best practices, phishing threats, and the importance of strong passwords can significantly enhance an organization's security posture. Furthermore, fostering a culture of security awareness ensures that the principles of the CIA Triad are not just technical guidelines but integral to an organization's operational ethos.

Expanding further into the nuances and evolving landscape of the CIA Triad in cybersecurity, let's explore the integration of emerging technologies, the significance of regulatory compliance, and the future trajectory of digital security measures.

CHAPTER 1   INTRODUCTION TO CYBERSECURITY

Integration of Emerging Technologies

Artificial Intelligence and Machine Learning

Artificial intelligence (AI) and machine learning (ML) are revolutionizing the approach to cybersecurity, impacting all three pillars of the CIA Triad. AI-driven systems can predict and detect cybersecurity threats in real-time, enhancing the integrity and confidentiality of data by preemptively identifying breach attempts before they occur. Moreover, AI algorithms optimize the allocation of resources, ensuring high availability by dynamically adjusting to demand and detecting potential system overloads or failures.

Internet of Things (IoT) Security

The proliferation of IoT devices introduces new challenges and complexities in maintaining the CIA Triad. Each device represents a potential entry point for attackers, making confidentiality and integrity harder to maintain. Simultaneously, these devices often require continuous uptime, emphasizing the importance of availability. Securing the IoT ecosystem involves robust encryption, regular software updates to maintain integrity, and resilient network architecture to ensure ongoing availability.

Regulatory Compliance and Data Protection

The importance of the CIA Triad is also underscored by the increasing number of regulations focused on data protection and privacy, such as the General Data Protection Regulation (GDPR) in the European Union and the California Consumer Privacy Act (CCPA) in the United States. These regulations mandate strict adherence to principles that align closely with confidentiality and integrity, requiring businesses to protect personal data against unauthorized access and ensuring accurate and timely processing. Compliance is not just a legal requirement but also a demonstration of an organization's commitment to cybersecurity best practices.

The Future Trajectory of Digital Security Measures

As digital transformation continues to evolve, so will the strategies to maintain the CIA Triad. Future security measures are likely to focus on even more sophisticated encryption techniques, advanced anomaly detection using AI, and the decentralization of data storage and processing to enhance security. Quantum computing presents both a challenge and an opportunity, potentially rendering current encryption methods obsolete while also offering new ways to secure data.

Collaboration and Sharing in Cybersecurity
The future of cybersecurity also lies in enhanced collaboration and information sharing between organizations, industries, and governments. By sharing threat intelligence and best practices, the global community can better anticipate and mitigate cyber threats, reinforcing the principles of the CIA Triad on a broader scale. This collaborative approach, combined with advancements in technology, will be crucial in staying ahead of increasingly sophisticated cyber threats.

Education and Training
Continued education and training in cybersecurity principles and practices are vital for all stakeholders, from IT professionals to end-users. As the human factor remains a critical vulnerability, increasing awareness and understanding of the importance of confidentiality, integrity, and availability can significantly reduce the risk of breaches. Through continuous learning and adaptation, organizations can foster a culture of security that supports the CIA Triad's goals.

Conclusion
The CIA Triad remains an essential framework for understanding and implementing cybersecurity measures, adapting over time to address new challenges, and integrating emerging technologies. By focusing on confidentiality, integrity, and availability, organizations can protect themselves against a wide range of cyber threats. However, it is the ongoing commitment to cybersecurity awareness, collaboration, and innovation that will ultimately determine the effectiveness of these efforts, ensuring a secure digital future for all.

CHAPTER 1   INTRODUCTION TO CYBERSECURITY

The Necessity of Protecting Data

In the digital age, our lives are increasingly documented, shared, and stored in the vast expanse of the digital world. This digital presence, while offering unparalleled convenience and connectivity, also presents a significant vulnerability: the risk to our personal data. The necessity of protecting this data cannot be overstated, as the implications of breaches touch every aspect of our lives, from our privacy and financial health to our very identities.

Privacy Concerns

The value of personal data in the digital marketplace has skyrocketed, with businesses and malicious actors alike recognizing its worth. Personal data, from the seemingly innocuous to the deeply personal, can paint a detailed picture of an individual's life, preferences, and behaviors. When these data fall into the wrong hands, the consequences can be profound.

Privacy, once breached, is difficult to reclaim. Unauthorized access to personal emails, messages, and documents can lead to embarrassment and psychological distress, while the exposure of sensitive financial or health records can have tangible, harmful effects. Beyond the individual, privacy breaches can erode the trust between consumers and companies, damaging relationships built on the assurance of data security. In extreme cases, breaches of personal data can even escalate to physical threats, as location data and personal routines are exploited by malicious actors.

This vulnerability underscores the critical need for robust cybersecurity measures to protect personal information. Encryption, secure password practices, and vigilant monitoring of personal data are fundamental to safeguarding privacy in the interconnected digital landscape.

Financial Implications

The financial repercussions of cybersecurity breaches extend far beyond the immediate theft of funds. For businesses, a data breach can dismantle the trust that took years to build with their customers, leading to a decline

CHAPTER 1   INTRODUCTION TO CYBERSECURITY

in sales and potentially devastating long-term financial consequences. The direct costs associated with addressing a breach — including legal fees, fines, and the expenses involved in securing the breach — can cripple a company's finances. Additionally, businesses often face increased insurance premiums and may need to invest significantly in upgrading their cybersecurity infrastructure to prevent future incidents.

Individuals are not spared from the financial fallout of data breaches. Victims of identity theft can spend years untangling the web of financial deceit spun with their stolen information. Restoring one's credit and regaining control over personal financial accounts can be a long, arduous process fraught with challenges. The emotional toll of dealing with identity theft, coupled with the potential for lasting financial insecurity, highlights the paramount importance of personal data protection.

The Potential for Identity Theft

Identity theft represents a particularly insidious threat in the digital age. Cybercriminals, armed with just a few pieces of personal information, can usurp an individual's identity to commit wide-ranging fraud. The internet's anonymity and complexity provide fertile ground for these thieves to operate, often leaving little trace until substantial damage has been done.

The methods employed by identity thieves are increasingly sophisticated, leveraging technology to harvest personal data at an alarming scale. Phishing attacks, which deceive individuals into divulging sensitive information, and breaches of corporate databases are among the tools in the cybercriminal's arsenal. The impact of identity theft on individuals can be catastrophic, affecting not just their financial health, but also their mental well-being and sense of security.

Protecting against identity theft requires a multifaceted approach, including both individual vigilance and corporate responsibility. It's essential for individuals to monitor their financial statements and personal data online, while businesses must secure the personal information they hold with the utmost care, employing state-of-the-art cybersecurity measures.

CHAPTER 1 INTRODUCTION TO CYBERSECURITY

Expanding upon the discussion of the necessity of protecting data with real-world examples illuminates the tangible impacts of cybersecurity breaches and identity theft. These instances serve as stark reminders of the vulnerabilities inherent in our digital ecosystem and underscore the critical importance of robust cybersecurity measures.

Real-World Examples of Privacy Breaches

1. **Equifax Data Breach (2017):** One of the most significant data breaches in history involved Equifax, a major credit reporting agency, which exposed the personal information of approximately 147 million people. Sensitive information, including Social Security numbers, birth dates, addresses, and in some instances, driver's license numbers, was compromised. This breach highlighted the devastating potential of attacks on data repositories and the long-term consequences for individual privacy and financial security.

2. **Yahoo Data Breach (2013-2014):** Yahoo experienced one of the largest data breaches ever recorded, affecting all 3 billion accounts on its platform. The breach included names, email addresses, telephone numbers, dates of birth, hashed passwords, and, in some cases, encrypted or unencrypted security questions and answers. This massive compromise of personal information demonstrated the extensive impact a single cybersecurity failure could have on global privacy.

CHAPTER 1   INTRODUCTION TO CYBERSECURITY

Financial Implications of Cybersecurity Breaches

1. **WannaCry Ransomware Attack (2017):** The WannaCry ransomware attack affected over 200,000 computers across 150 countries, with total damage ranging from hundreds of millions to billions of dollars. Critical systems, including those in healthcare, government, and finance, were encrypted by the attack, demanding ransom payments for data restoration. The financial implications of this attack were monumental, disrupting services and causing significant operational and recovery costs.

2. **Target Data Breach (2013):** Target, the American retail giant, suffered a data breach in which attackers accessed approximately 40 million credit and debit card numbers. The breach not only led to direct financial losses for the company, estimated at $162 million, but also significantly damaged Target's reputation, affecting sales and customer trust for years following the incident.

The Potential for Identity Theft

1. **IRS Identity Theft (2015):** Cybercriminals exploited a vulnerability in the IRS's "Get Transcript" application to gain unauthorized access to approximately 700,000 taxpayer accounts. By using previously stolen personal information, attackers filed fraudulent tax returns, diverting refunds and causing immense financial and administrative distress for the victims.

2. **Anthem Data Breach (2015):** In one of the largest healthcare data breaches, hackers accessed the personal information of nearly 80 million Anthem insurance customers. The exposed data included names, birthdays, medical IDs, Social Security numbers, addresses, emails, employment information, and income data, leaving millions vulnerable to identity theft and fraud.

In Reflection
These examples illustrate the diverse and profound impacts of cybersecurity breaches on privacy, financial well-being, and the potential for identity theft. They serve as a clarion call for heightened vigilance and robust security measures to protect personal and organizational data. Beyond the immediate repercussions for affected individuals and companies, these breaches have broader implications for society's trust in digital systems and the integrity of our interconnected world.

As we continue to forge ahead in the digital age, these real-world examples remind us of the ever-present need to prioritize cybersecurity, not only as a means of protection but also as a foundation for trust and confidence in the digital landscape.

Reflecting on these impactful cybersecurity incidents leads to a deeper understanding of the multifaceted challenges we face in protecting digital data. These real-world examples underscore the necessity for comprehensive cybersecurity strategies that address not only the technological vulnerabilities but also the human elements that often serve as the weakest links in security chains.

Expanding the Conversation: Lessons Learned
The aftermath of these incidents brings to light several critical lessons for both organizations and individuals. First, the importance of proactive cybersecurity measures — such as regular software updates, robust encryption practices, and advanced threat detection systems — cannot

be overstated. Equally important is the need for ongoing cybersecurity education and awareness programs to equip all users with the knowledge to identify and avoid potential threats.

Furthermore, these examples highlight the need for strong incident response plans. The ability of an organization to quickly detect, respond to, and recover from a breach can significantly mitigate its impacts. This includes transparent communication with affected parties and regulatory bodies, which is crucial for maintaining trust and accountability.

Strengthening Cybersecurity Frameworks

In response to growing cyber threats, governments and international organizations have developed regulations and standards to strengthen the cybersecurity posture of organizations. GDPR (General Data Protection Regulation) in the European Union and CCPA (California Consumer Privacy Act) in the United States are examples of legislative efforts aimed at protecting personal data. Compliance with such regulations helps not only in safeguarding data but also in building consumer confidence.

Industry standards like ISO/IEC 27001 provide a framework for managing information security. They offer best practices on risk management, security controls, and compliance, helping organizations of all sizes and sectors to fortify their cybersecurity defenses.

The Role of Emerging Technologies

Emerging technologies also play a pivotal role in advancing cybersecurity. Artificial intelligence (AI) and machine learning (ML) are being leveraged to predict and identify cyber threats more efficiently. Blockchain technology offers new ways to secure transactions and data integrity through its decentralized and tamper-evident nature.

However, as these technologies evolve, so do the tactics of cybercriminals. The dynamic nature of cybersecurity necessitates a continuous adaptation and evolution of security strategies to stay ahead of threats.

Collective Responsibility and Collaboration

The fight against cyber threats is not the sole responsibility of cybersecurity professionals or IT departments; it is a collective responsibility that requires the participation of all stakeholders, including users, organizations, and governments. Public-private partnerships and information sharing between entities play a critical role in identifying and mitigating threats on a global scale.

Cybersecurity alliances and forums facilitate the exchange of threat intelligence and best practices, enhancing the collective ability to respond to and prevent attacks. Such collaboration is essential for developing a resilient digital ecosystem capable of withstanding the evolving landscape of cyber threats.

Conclusion

The journey through the landscape of cybersecurity is complex and fraught with challenges. Yet, it is also an opportunity for innovation, collaboration, and advancement in the ways we protect and secure our digital world. As we reflect on the real-world examples of data breaches and the lessons they impart, it becomes clear that cybersecurity is not just about defending against attacks — it's about fostering a secure, trustworthy, and resilient digital environment for all.

In conclusion, the necessity of protecting data in the digital age is a multifaceted challenge that demands a comprehensive and adaptive approach. From implementing advanced technological defenses to fostering a culture of security awareness, the efforts to safeguard our digital information are critical in navigating the complexities of the modern world. By learning from past incidents and embracing a collaborative stance toward cybersecurity, we can aspire to a future where digital innovations flourish without compromising the security and privacy of our data.

References

[1] Stallings, W. (2017). Cryptography and Network Security: Principles and Practice. Pearson Education, Inc. This book provides an in-depth look at the cryptographic techniques that underpin the confidentiality and integrity of data.

[2] Whitman, M. E., & Mattord, H. J. (2018). Principles of Information Security. Cengage Learning. This text explores the foundational principles of information security, including the CIA triad, and their application in protecting digital information.

[3] Peltier, T. R. (2016). Information Security Policies, Procedures, and Standards: Guidelines for Effective Information Security Management. Auerbach Publications. This book discusses the importance of organizational policies and procedures in maintaining the confidentiality, integrity, and availability of information.

[4] European Parliament and Council of the European Union. (2016). Regulation (EU) 2016/679 of the European Parliament and of the Council of 27 April 2016 on the protection of natural persons with regard to the processing of personal data and on the free movement of such data (General Data Protection Regulation). Official Journal of the European Union.

[5] California Legislative Information. (2018). California Consumer Privacy Act (CCPA). Senate Bill No. 1121.

[6] Romanosky, S. (2016). Examining the costs and causes of cyber incidents. Journal of Cybersecurity, 2(2), 121-135.

CHAPTER 2

Threat Landscape

Author:
Anirudh Khanna

Overview of Current and Emerging Threats

With the progressive development of the digital landscape, threats are continually evolving, leading to several challenges in technological applications. Threats have developed based on their capacity to affect cyber systems, necessitating understanding the threats and providing remarkable steps to mitigate and prevent them reasonably.

CHAPTER 2   THREAT LANDSCAPE

Current Threats

Phishing and Social Engineering

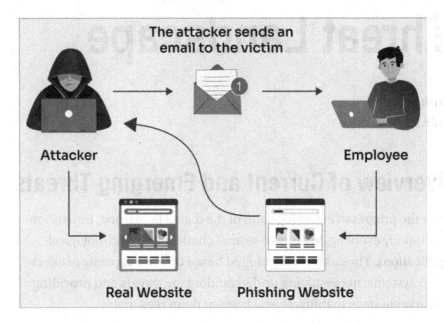

Figure 2-1. *Process of Phishing*

Figure 2-1 highlights critical information about phishing as it occurs. This indicates the primary technique: the attacker clones a real website, sends a deceptive email to the victim with a link to the fake website, and collects information from the fake website when employees log in. This image details the nature of phishing and its occurrence among unsuspecting employees.

Phishing and social engineering are common and persistent threats occurring in the digital landscape. These two techniques ensure the utilization of human psychology to deceive victims and let them reveal sensitive information that leads to access to systems or gaining data for use in different activities. Primarily, phishing is when the attackers use the guise of trustworthy parties to trick victims into revealing sensitive

CHAPTER 2   THREAT LANDSCAPE

information. These attackers demand information such as credit card numbers, usernames, passwords, and key personal details. Different phishing attacks must be skillfully handled to achieve desirable modeling of threats as they occur in any instance [1]. Phishing uses different techniques to ensure they can gather information from victims. The use of urgency and fear tactics indicates to victims that they have to address an urgent issue to enable immediate action. Attackers also use emails and websites that resemble legitimate ones, making it more straightforward to gather the information. Brand imitation is also used to ensure that they can mimic company logos and have a close appeal to the company to target employees and customers. There are different kinds of phishing attacks, depending on the approach created by a person to help them cater to an activity and ensure they address specific needs. These kinds of phishing attacks include

A. Email Phishing

   This is a technique where attackers send deceptive emails to the victims. These emails come from reputable sources, and they ensure they can communicate with the intended party with essential instructions, leading them to open specific links or attachments to the email. This indicates that they must create the right channel to enable the victim to trust their emails and showcase the trust by clicking the link or attachment, leading to a false website that would help collect their information [2]. Therefore, this process takes advantage of the victims and leads them to provide sensitive information, helping the attackers access various sites. Notably, email phishing ensures that there is mass distribution of the information across a large number of target clients. They use language that will become

CHAPTER 2 THREAT LANDSCAPE

deceptive to a particular audience, leading them to click on the provided links. Email phishing has an aggressive tendency since attackers can continually send messages that place them in a position to ensure they can gather information and outline every activity they require the victims to do. Using this approach, email phishing gains prominence in spamming users and is often assumed by most users. User awareness and education about phishing patterns have helped safeguard against more attacks.

B. Spear Phishing

This is a specific targeting of victims where they consider personal information and victim outlines to help them make a more credible appeal for the victim. Spear phishing involves emails that have the victim's information, where they consider their roles and responsibilities, helping them understand the instructions, which leads to revealing sensitive information. The model enables the use of the approach to ensure that victims can reveal their information on demand since they believe that the emails are from reliable sources. Notably, the spear phishing model ensures the accurate development and depiction of approaches seeking to target an individual, implying that attackers must generate prior knowledge about the party and ensure they learn more about it. Information used for spear phishing has to be personalized to a level where it attends to every requirement and consideration of the victim, intending to make every element of the message believable and considered legitimate to achieve the key information provided by the parties. Thus, attackers have to learn about every communication model between the company and the victim, making it much easier to establish a legitimate connection for their actions. Spear phishing has high stakes because of the amount of time taken to help gather information and employ the proper techniques to gain trust and exploit the victim. These stakes relate to having full-scale attacks later on or having the entire client information used for other attacks to help advance their demands.

C. Whaling

This type of phishing mainly targets executives and senior officials within the government. The emails require urgent action from the parties, forcing them to act without considering the activities. Phishing mainly asks for money transfers to help address and mark the development of a new way to handle information promptly.

D. Smishing

This model of phishing employs SMS to ensure that it targets individuals and conveys instructions that they must follow to gain sensitive details. These SMSs can have prompts that force the subjects to reveal their course of action for urgent action and even urge the victims to click on a link to help resolve an issue with their financial accounts.

CHAPTER 2   THREAT LANDSCAPE

E. Vishing

This model uses voices to convince victims to reveal sensitive information. The approach enables the attacker and the instructor to have direct engagement with victims, making it easier to reveal information and even dwell on activities that should be done to receive their sensitive information.

More to the point, social engineering, as an aspect of phishing, works by deceiving victims to carry out activities through techniques that lead them to carry out these activities with much ease. Social engineering can be conducted in different ways, ensuring that the attackers have a significant benefit over victims and gathering the said information from them. Fundamental techniques in social engineering include

A. Baiting

This process works when an attacker uses a technique to lure the victims. For instance, the attackers might drop a USB labeled confidential, hoping the victim will plug into their device, wanting to see the information on the device and, in turn, using it to gain information from their devices. Baiting entirely depends on the victim's action, ensuring they can engage in several activities that define and appeal to their critical appeal in managing different communication engagements. The baiting approach has to close it or the e-target and ensure they can be studied to learn about their concerns when executing the roach.

CHAPTER 2 THREAT LANDSCAPE

B. Pretexting

This technique happens when the attacker creates a scenario to help them gather information from the target. This mechanism ensures they can access sensitive information or get the victims to perform certain activities.

C. Impersonation

In this case, the attacker gains the victim's trust, helping them to collect sensitive information for their deeds.

D. Tailgating

In this instance, the attackers gain access by following the victim and ensuring they have information in a restricted area without the victim's knowledge [3].

E. Quid Pro Quo

This involves the attacker offering something in exchange of information and sensitive data to help them gain access. This approach ensures that the attacker appears as someone needing help but can also gather assistance for the victim through certain favors.

Ransomware

This type of malware refuses a victim access to their data until they pay an amount of money, usually in bitcoins, to the attacker. Most often, there is no guarantee of providing access to data even after ransom payment [4]. Ransomware has evolved over the ages, continually taking advantage of users and ensuring they can provide payment so their information is returned to them.

CHAPTER 2   THREAT LANDSCAPE

Figure 2-2. *Process of a ransomware attack*

Figure 2-2 highlights the critical procedures of a ransomware attack. The processes begin with receiving malware, followed by its deployment on the victim's computer system. The process follows with decrypting files on the computer and displaying a notice for ransom, where the victim is expected to pay to gain access to the files. This approach completes the ransomware process, ensuring that every essential action maintains the attacker's leverage.

Malware

Different kinds of malicious software can be used to ensure that they disrupt damage or even provide computer entry. Common kinds of malware include

I. **Viruses**: These malicious codes come alongside legitimate programs and spread when the program is executed on the computer system.

II. **Trojans**: This is malware disguised as software that prompts users to install, allowing the attackers to gain access.

III. Worms are malicious software that spreads even without action from the victim; they often spread through network connections.

IV. **Adware**: This malware downloads and displays intrusive and unwanted advertisements to the users.

V. **Spyware**: This malicious material monitors user activity, collets activity, and information from the platform.

VI. Rootkits are tools that enable attackers to gain and maintain access to the system without being noticed.

Botnets

These are a network of infected computers. Attackers control these computers and help them carry out different malicious activities, such as spamming, DDOS attacks, and spreading malware. The attackers must use several devices to ensure increased leverage in launching their attack. The process could go on for a prolonged period, leading to massive damage and the handling of information.

Advanced Persistent Threats (APTs)

These are cyberattacks where the intruder gains access to the system. In these attacks, the attacker is undetected for a long time, and they can monitor and collect information on the system for their use. These attacks are often targeted and specific for the users, increasing their capacity to handle their intentions. In most cases, they are conducted to assist with espionage and sabotage activities [5].

Zero-Day Exploits

These attacks look for unknown vulnerabilities and weaknesses in software and hardware. The attackers ensure that they leverage the lack of defenses and work to ensure they can gain as much information as possible to achieve a suitable appeal. These zero-day attacks enable attackers to perform their activities and complete them without being noticed as ever present on the system.

CHAPTER 2   THREAT LANDSCAPE

Emerging Threats

The nature of threats constantly keeps evolving to become more sophisticated and have diverse approaches that impact their handling. These threats emerge as they seek to address existing mitigation strategies and new technological advances that come with new weaknesses and vulnerabilities to the exploited. The emerging threats also bring out a demand to ensure users understand these techniques and update security models that assist them with proper protection [6]. Most importantly, these emerging threats are built to bypass the traditional and existing security approaches, ensuring that there are easy ways to gain access to systems and manipulate them to have an easy application in whatever way is needed. The lack of sufficient defense mechanisms against these platforms leads to a high possibility of success in achieving their objectives at any level.

Artificial Intelligence (AI)-Driven Attacks

The development of AI has aided the creation of ways to ensure that it assists in enhancing the effectiveness and accuracy of cyberattacks. Artificial intelligence helps automate phishing and leads to models that will help address constant development to achieve the required intention. The automation and improvement of phishing through AI have increased the possibility of success in managing attacks and affecting different systems. AI-driven attacks exist in different dimensions, leading to a remarkable level of attacks being administered to users and affecting the scope and capacity to achieve the goals demanded. Some key examples of these attacks include

 I. **Adversarial Machine Learning**: This involves the manipulation of inputs on AI systems to ensure they make incorrect predictions. Attackers create systems to ensure that they can manipulate AI to handle activities in whatever way they deem possible.

II. **Data Poisoning**: This approach corrupts data used to train different machine learning models. The process ensures that the machines cannot conduct their intended duties but do as the attacker demands [7].

III. **Automated Phishing and Social Engineering**: Training AI to conduct social engineering and phishing activities to enhance their scale of operation and accuracy of providing the information to achieve the objectives.

IV. **Deep Fake Attacks**: Using AI to create fake videos and audio to ensure they manipulate individuals. This ensures that they can be used even in social engineering to target individuals and lead to a collection of sensitive information.

V. **Model Inversion and Extraction**: Attackers use this approach to ensure they can collect personally identifying information while conducting machine learning training activities. The threat lies in identifying the information, which can lead to breaches within the system.

VI. **AI-Powered Malware**: It enables attackers to have more innovative malware that adapts its behavior to different threat detection tools, making it unrecognizable in most instances. In this case, AI empowers the malware to adapt and cause more damage to the system.

VII. **Automated Vulnerability Discovery**: This is a mechanism where attackers train AI to help them assess systems and investigate vulnerabilities. Exploiting AI systems' accuracy allows attackers to ensure they can identify weaknesses and investigate potential entry points within these systems. The approach also allows attackers to ensure that it can handle every model approach and achieve sustainable handling of the attack to achieve a higher outcome [8].

Internet of Things (IoT) Vulnerabilities

The increasing number of connected devices in modern companies leads to an increased threat level. The IoT devices have a high level of influence and impact on the system to achieve a suitable scale and level of attack handling. Moreover, IoT devices come with weak security mechanisms, making it easier for attackers to enter the system through the IoT devices in a home or office. Thus, these devices make an easy target and entry point into the network, offering more points of vulnerability to be exploited.

Cloud Security Threats

More companies are running their services on cloud computing platforms. Companies have high storage capacities on cloud networks, leading to a high chance of vulnerability when addressing their functions. However, emerging threats lie within breaches on the cloud platforms when users do not understand the correct settings to govern their information. This approach indicates that the vulnerabilities on the cloud platform lie within the capacity and scope of ensuring an increasingly beneficial point of administering their values and needs to achieve the right outcome. Thus, with misconfigured settings, cloud security poses a higher threat to users as attackers can exploit these vulnerabilities to gather data from the systems.

Quantum Computing Threats

Quantum computing poses a significant risk in addressing encryption and uses by attackers to gain leverage. Attackers could use the computational power of quantum computing to decrypt encryptions on sensitive information. This poses a high risk to cybersecurity in various cloud security channels, often affecting the potential to address and monitor their engagements to achieve a particular outcome.

Mitigation Strategies

Different strategies can be used to ensure increased security in addressing and ensuring suitable management of the safety mechanisms of organizations. These current and emerging threats must be handled in critical ways that assist in addressing significant challenges and channels of security within cyberspace. Companies must use these strategic additions to safeguard and work within their required levels to achieve the most remarkable level of advancing value to the right angle. Thus, working within these dimensions and capacities to achieve the right level of safety demands that every engagement be conducted to achieve the most valuable adjustment in all instances. Thus, the proper mitigation strategies must ensure an increasingly beneficial way to attract value to handling cyber threats. Vital strategic activities include:

A. **AI-Powered Defense Systems**: These are systems used to address continued assessment of the cybersecurity defense and provide alerts on detecting anomalies. In this case, they are vital in addressing risks and identifying major approaches to ensure that there are better ways to respond to threats. Automation can also help assess any issue and lead to better handling of threats before they occur.

B. **Data Security and Privacy Measures:** Organizations must ensure that they establish strong policies and regulations to safeguard data. The right politics will impact data governance and handling in the company, leading to a remarkable level of addressing and managing the information handling approach.

C. **Adversarial Training:** Engaging adversarial techniques will enhance the robustness of training models to fight against attacks of the same kind. This will enable a structured level of dwelling on pertinent challenges affecting the AI training models.

D. **User Education and Awareness Campaigns:** Enlightening users, from employees to customers, about imminent threats and techniques used by attackers provides a chance to learn about critical issues they must tackle to achieve the best results [9]. User education and awareness approaches will enable a considerable development of knowledge on core issues, providing a remarkable perspective in tackling the threats that affect the system.

E. **Regular Assessment and Penetration Testing:** Companies must conduct regular audits and identify vulnerabilities within the system to assist them in handling the current challenges. This mechanism will ensure that companies know their weak points and can consistently address their handling to achieve a remarkable outcome.

CHAPTER 2 THREAT LANDSCAPE

F. **Collaboration and Information Sharing**: This is a crucial step to ensure different engagements between stakeholders, government agencies, and experts to constantly gain information on best practices to address the challenge of working with imminent threats to the system.

G. **Regulatory Compliance**: Complying with regulatory needs helps ensure the organization is updated on protection mechanisms to provide a critical appeal to every evolving threat. Regulations specific to industries must be addressed to ensure companies can conduct their activities more quickly.

H. **Continual Monitoring and Adaptation**: Companies must continually assess their capacity to handle cyber threats and work within the steps needed to achieve remarkable results. These advances have to be made to adapt to the changing nature of attacks and include new strategies to mitigate against threats to their systems [10]. Hence, the approach is a key step in enabling companies to continue being informed about every activity they must carry out.

Cybercrime, Malware, and State-Sponsored Attacks

Cybercrime

Cybercrime comprises illegal activities conducted on digital platforms for various intentions. Cybercriminals are often motivated to gain financial gain from their victims. The process ensures profit using different mechanisms to steal money from unsuspecting victims. These parties' approach often leads them to steal personal data, money, and intellectual property.

In most instances, cybercriminals operate in groups that help them leverage their skills and competencies, enabling them to have the right approach to handling any upcoming issues. Cybercriminal organizations are often made to ensure that they can address several needs and activities, each aimed at enabling a progressive handling and management of their operations. Some key cybercriminal activities include identity theft, ransomware, phishing, and credit card fraud.

The primary motivation of financial gain for credit card fraud enables criminals to gather data, sell it for financial gain, and ensure they can look for more corporate, personal, or intellectual data to sell on the dark net for their profiteering purposes [11]. During ransomware, phishing, and credit card fraud, these criminals use the information to gain money from their victims. In some instances, extortion and blackmail help to gain money from the victims when they threaten to release information from the collected data. In this case, these approaches ensure that cybercriminal activities are monetized to help with financial gain and leverage the attacker's needs to gain more money.

Malware Campaigns

Malware is software designed to disrupt, damage, and gain unauthorized system access. It can be used for several purposes, such as enabling financial gain and sabotaging or espionage. Different kinds of malware can be deployed to address the attacker's various needs. Common types of malware include viruses, adware, Trojans, and spyware.

Most malware campaigns are motivated by different appeals aimed at securing and appealing to different demands. The key motivations include

1. **Data Theft and Espionage**: Malware campaigns can steal sensitive information for different activities. Malware also applies to handling geopolitical and business differences, where information is stolen to assist business entities in gaining a competitive advantage over others. Moreover, a geopolitical malware campaign steals information to address policies and craft developments in nations.

2. **Disruption and Damage**: Malware campaigns can be used to ensure that no activities are running for business entities. This motivation purely excludes the demand to gain financial gain from the activities. It concentrates on the need to identify and work within the provision of a step to damage operations for an entity at any given time. In some cases, destructive malware is used to render files unusable, leading to a challenge for companies to fail in handling their needs at any given point [12].

3. **Control and Manipulation**: Malware can be used to collect information and gain control of systems to ensure they manipulate operations in a desired manner. Attackers can use the malware to ensure that they have appropriate control, which helps them address their own needs within the system, further affecting the functionality and possibility of achieving intended needs. In this sense, the use of the malware affects the chance to provide and detail appropriate modeling and management of the underlying needs within the system.

4. **Monetization**: Malware campaigns can be used to gain more revenue for attackers further when they manipulate the systems. These malware campaigns can earn more revenue by forcing ad clicks, leading to many individuals clicking on ads and generating revenue for the attackers without their consent and knowledge [13].

State-Sponsored Cyber Warfare

These are attacks that are done on behalf of nation-states. They are meant to achieve objectives that align with the demand to ensure these nations are ahead in their plans against enemies or friendly nations. The attacks often target critical infrastructure, private sector organizations, and military assets. The attacks are often political, ideological, and strategic in gaining the suitable capacity to address national needs. These attacks are well-resourced and sophisticated, leading to a high possibility of addressing their needs without fail. They commonly use zero-day exploits, allowing the nations to have unrecognized access to the system, enabling them to continually handle their needs and achieve the desired outcome at whatever rate. Therefore, the approach creates a mechanism where state-sponsored attacks have a highly sophisticated appeal, lacking the right opportunity to target and address their underlying needs [14]. Therefore, the cyber operations of the nations are meant to primarily lead to a high level of data monitoring and analysis within these nations to help gain leverage as needed.

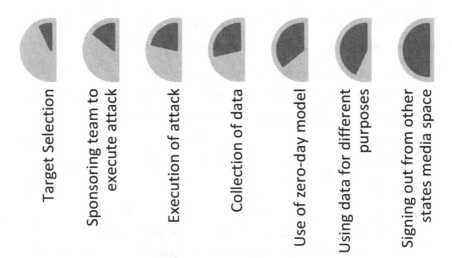

Figure 2-3. *Process of State-Sponsored Attacks*

Figure 2-3 highlights the critical processes for use in addressing state-sponsored attacks. These processes begin with target selection, where nation-states select the infrastructure, amenity, and country they want to attack. The following process is selecting and sponsoring a team to help them execute the attack. The next step is executing the attack, mainly conducted through the zero-day approach. Data collection is a critical step in ensuring that information is collected and addressed to ensure the proper modeling of every demand for the nation-state. Zero-day attacks are used to ensure that they can stay undetected on the platform, enabling the use of the information for different processes. The nation-state attacks end with the sponsored party getting out of the platform to ensure they can use the collected information for a meaningful outcome.

CHAPTER 2   THREAT LANDSCAPE

These attacks have different motivations to create the best outcome for sponsoring nations. The key motivations include

- A. **Government Secrets**: These attacks could be used to ensure that they gain access to sensitive communication within the government. The approach also ensures access to classified information in the government, enhancing the perception and understanding of every government activity.

- B. **Espionage**: The attacks can help monitor nations and gather intelligence on critical points within political and business entities to help address strategic advantages. The process leads to a high level of addressing organizational and national capacity to gather intelligence about the target entity, leading to better and higher outcomes in defining their activities to have a desired appeal.

- C. **Industrial Espionage**: This can be used on international companies to ensure an excellent understanding of their strategic plans and operations within the company. The espionage runs long, enabling an understanding of activities that can lead to a greater identity and definition of activities as they should be occurring within the nations.

- D. **Political and Strategic Motivation**: This can be conducted to ensure geopolitical advancement, ensuring that they can manipulate the countries based on their decisions and have secret information helping them to advance and achieve objectives that can otherwise lead nations to always stay ahead of their allies or competitor nations.

E. **Cyber Deterrence and Warfare:** This can be conducted on nations to ensure that they turn off the capabilities of enemy nations and prevent them from having the suitable capacity to fight back or address particular functionalities. The motivation also provides that a nation can show its cyber prowess, account for advancing needs and channels, and ensure sustainable and appropriate management of every core adjustment to achieve the desired outcome [15].

F. **Sabotage and Disruption:** Nations can sponsor long-term attacks on adversaries to help them learn about their critical infrastructure and military models. This attack begins with monitoring their functionalities and investigating activities they can conduct to help weaken the adversary. The cyberattack is motivated by the need to sabotage transport systems, have a weak communication network, and even have a weak power grid, always affecting activities within the country. These activities indicate a high chance of influencing nations to have weak systems because of intentional and well-crafted cyberattacks to have the desired meaning.

Figure 2-4. *Motivations for State-Sponsored Cyber Warfare*

Figure 2-4 highlights various motivations for state-sponsored cyber warfare activities. These reasons include sabotage and disruption, cyber deterrence and warfare promotion, political and strategic motivation, industrial espionage, and exposure of government secrets. These motivations detail demands by state nations to collude with entities to promote cyber warfare and ensure sponsoring malware campaigns that can span an extended period.

Distinguishing Between Cybercrime and StateSsponsored Cyber Warfare

In most instances, cybercrime, state-sponsored attacks, and malware campaigns have critical distinctions in the form of being executed and handled. However, there are certain instances where they overlap and could have a similar outfit. There are instances when nation-states can collaborate with cybercriminal organizations to attack different parties. The attacks conducted on international organizations, other states, and critical infrastructure have blurred the line in defining and addressing each

CHAPTER 2 THREAT LANDSCAPE

cyberattack. Additionally, groups can copy techniques from one another, making them have difficulty attributing a method of attack on another group. This implies that the entire engagement might lead to a more significant demand to identify and look into critical approaches that have to be conducted to lead to significant distinctions in managing the attacks and their forms of appeal.

On the other hand, the motivations are critical as they help determine the force behind an attack. More to the point, addressing the motivations helps to craft the most effective defenses that can be used to address and administer a reliable way to conduct a more profound handling of cybersecurity. Organizations and entities that are imminent to attack must tailor their strategies to address and work within the correct dimension of addressing every underlying need to achieve remarkable solutions. Therefore, properly understanding and addressing the cybersecurity strategies will help companies stand against any upcoming challenges, ensuring they can schedule and administer valuable outcomes to whatever rightful entity is demanded of them. Thus, these advances are key to considering the best option in administering valuable engagement of the cybersecurity mentions.

Cybercrime and state-sponsored attacks also have different natures of actions and targets. Cybercriminal networks concentrate on the possibility of ensuring that they launch sophisticated attacks that target financial gains from an individual or business entities by getting into their online channels or financial accounts. Nonetheless, state-sponsored cyber warfare manages highly sophisticated systems aimed at the target states' military and critical infrastructure. This indicates the distinct demand to ensure remarkable handling and address several ways to understand the two cyber threats in modern society.

Legally, these threats have different ramifications. The use of cybercrime is subject to law enforcement and judicial process. Different law enforcement parties can ensure that they present the right level of understanding and address the issue to ensure that they can continue engaging both parties and have the right form of communication about the underlying problem. The legal ramifications include fines or prison terms for the perpetrators of cybercrimes since they are chargeable by law. On the other hand, the nations take state-sponsored cyber warfare as an act of aggression. Even though they have lasting influences on the nations and critical infrastructure under attack, they only have international diplomatic issues and retaliatory attacks. There are no legal ramifications for the state-sponsored attacks. However, nations have over the years developed approaches seeking to ensure that they deal with any upcoming attacks, providing strategies to ensure that they can retaliate and communicate about every underlying issue that might affect the provision of safe modeling of their cyberinfrastructure to suitably handle the needs of security management.

Summary

Given the rapidly evolving nature of technology, cyber threats also constantly evolve, remarking a demand to understand and address significant steps in marking positive development of protection against these threats. Critical threats include phishing and social engineering. Phishing techniques such as whaling, smishing, email phishing, and spear phishing are widely used to help address attacker demands to gather information. Social engineering models such as pretexting, impersonation, and tailgating also assist attackers in gathering quick information from websites, granting them the capacity to have information they can use for different gains. Further threats include ransomware, malware, APTs, and botnets. With the growing comprehension of technology, emerging threats like AI-powered threats, cloud computing threats, and IoT vulnerabilities must be addressed to enable secure management of threat addresses in organizations. These threats can be managed through regular system monitoring and updates, user awareness, and AI-powered defense systems to help instill excellent security in the network systems. More to the point,

CHAPTER 2 THREAT LANDSCAPE

cybercrime occurs through different individual attacks or those pushed by groups seeking to infiltrate information. Cybercrime is motivated by the need for monetization, unlike state-sponsored warfare, which is motivated by the need to obtain government secrets and enforce a cyber deterrence action on the websites. These factors must be keenly engaged in providing robust protection for cyber systems, mainly since there is an overlap between state-sponsored cyberattacks and cybercrime. Security protocols are vital to addressing imminent and current threats in the cyber universe.

References

[1] Z. Alkhalil, C. Hewage, L. Nawaf, and I. Khan, "Phishing attacks: A recent comprehensive study and a new anatomy," *Frontiers in Computer Science*, vol. 3, p. 563060, 2021.

[2] R. Alabdan, "Phishing attacks survey: Types, vectors, and technical approaches," *Future Internet*, vol. 12, no. 10, p. 168, 2020.

[3] K. Chetioui, B. Bah, A. O. Alami, and A. Bahnasse, "Overview of social engineering attacks on social networks," *Procedia Computer Science*, vol. 198, pp. 656-661, 2022.

[4] H. Oz, A. Aris, A. Levi, and A. S. Uluagac, "A survey on ransomware: Evolution, taxonomy, and defense solutions," *ACM Computing Surveys* (CSUR), vol. 54, no. 11s, pp. 1-37, 2022.

[5] H. Owen, J. Zarrin, and S. M. Pour, "A survey on botnets, issues, threats, methods, detection and prevention," *Journal of Cybersecurity and Privacy*, vol. 2, no. 1, pp. 74-88, 2022.

[6] D. Ghelani, "Cyber security, cyber threats, implications and future perspectives: A review," *Authorea Preprints*, 2022.

[7] B. Guembe, A. Azeta, S. Misra, V. C. Osamor, L. Fernandez-Sanz, and V. Pospelova, "The emerging threat of AI-driven cyber attacks: A review," *Applied Artificial Intelligence*, vol. 36, no. 1, p. 2037254, 2022.

[8] N. Kaloudi and J. Li, "The AI-based cyber threat landscape: A survey," *ACM Computing Surveys (CSUR)*, vol. 53, no. 1, pp. 1–34, 2020.

[9] H. Riggs et al., "Impact, vulnerabilities, and mitigation strategies for cyber-secure critical infrastructure," *Sensors*, vol. 23, no. 8, p. 4060, 2023.

[10] M. Ahsan, K. E. Nygard, R. Gomes, M. M. Chowdhury, N. Rifat, and J. F. Connolly, "Cybersecurity threats and their mitigation approaches using machine learning—A review," *Journal of Cybersecurity and Privacy*, vol. 2, no. 3, pp. 527–555, 2022.

[11] S. Chng, H. Y. Lu, A. Kumar, and D. Yau, "Hacker types, motivations and strategies: A comprehensive framework," *Computers in Human Behavior Reports*, vol. 5, p. 100167, 2022.

[12] G. M. Chiong, "The rise of ransomware: Motivations, contributing factors, and defenses," Ph.D. dissertation, Utica University, 2023.

[13] N. Brubaker, D. K. Zafra, K. Lunden, K. Proska, and C. Hildebrandt, "Financially motivated actors are expanding access into OT: Analysis of kill lists that include OT processes used with seven malware families," 2020.

[14] L. Burita and D. T. Le, "Cyber security and APT groups," in *2021 Communication and Information Technologies* (KIT), 2021, pp. 1–7.

[15] T. Miller, A. Staves, S. Maesschalck, M. Sturdee, and B. Green, "Looking back to look forward: Lessons learned from cyber-attacks on industrial control systems," *International Journal of Critical Infrastructure Protection*, vol. 35, p. 100464, 2021.

CHAPTER 3

Security Principles

Author:
Anirudh Khanna

Detailed Exploration of Encryption

Encryption is a fundamental requirement in the modern cyber society. Encryption ensures data protection by converting it to a format called ciphertext. The ciphertext is an unreadable format that demands deciphering before the information can be available to any party to read and understand. The deciphering of the ciphertext occurs only when an authorized party has the correct decryption key that helps them convert the document back to a readable format [1]. This indicates that encryption is critical in advancing safety by making data inaccessible to unauthorized parties, detailing its vital role in providing sustainable privacy and protection of sensitive information to ensure remarkable development in targeting and achieving the proper conveyance of communication.

Types of Encryptions

Two major types of encryptions provide a great insight into the model and nature of addressing tasks that require encryption. Encryption can be symmetric or asymmetric, making them the main types of encryptions widely used in contemporary industries.

Symmetric Encryption

Symmetric encryption is a model that uses the same key for encryption and decryption. This implies that the single key has great power over the information and must keep it secret between the parties. Since the model uses a single key to encrypt and decrypt, it is simpler to apply and use, making most companies use it to handle their secret information. This indicates the capacity to ensure that the key carries much information about the data.

This mechanism generates a secret key, after which plaintext information is converted using the secret key, making it a secure ciphertext that anyone cannot access without the right key. The information is then sent to the receiver, who can use the same key to ensure that they can convert it back to plaintext, which enhances their address of the information and access to whatever kind of information they are working on [2]. Thus, this approach enables suitable handling and management of the approach to deal with the modeling of information to be addressed selectively. Anyone with the secret key has access to the information, implying that the security is based on the capacity of a person holding the secret key as a secret that enables them to understand and work on the provided information.

Different kinds of encryption models use symmetric algorithms to help them address and manage the management and handling of the information. The Advanced encryption standard (AES) uses the mechanism and is widely used in government entities seeking to achieve a significant step in the safety and security of information. Additionally, the use of data encryption standard (DES) is also a symmetric model, even though AES has replaced it to enable continued service and management of information on the different platforms [3]. DES has been enhanced to have a triple appeal, ensuring sensible and critical management of the information to address key concerns associated with managing the generation of keys and addressing influential steps to protect the data.

CHAPTER 3 SECURITY PRINCIPLES

Symmetric encryption is used in different instances, helping to ensure data privacy and safety at all times. Primarily, the model assists with data storage, ensuring that the files and databases have to be stored while protecting sensitive information, which can be influential in marking the development of critical concerns associated with data handling to different parties. The model of encryption also works with communication, where they secure transmission of information on networks through virtual private networks (VPNs) and TLS (transport layer security) to assist with web traffic. Device security can also be configured using symmetric encryption, ensuring an instrumental capacity to prevent unauthorized access to devices whenever they have been stolen or lost. This indicates that the encryption model has several modern uses, encouraging continued management and handling to achieve remarkable results and the best outcome in addressing individual demands at all times.

Figure 3-1. *Symmetric Encryption Model of Action*

Figure 3-1 indicates how symmetric encryption works to ensure that plaintext is changed to ciphertext. The image suggests using a secret key to assist in changing plaintext to ciphertext and back during encryption. The image dictates using the same secret key in encryption and decryption.

CHAPTER 3   SECURITY PRINCIPLES

Asymmetric Encryption

This model of encryption is also referred to as public-key encryption. Asymmetric encryption uses two sets of keys, public and private, to ensure that they can protect data and prevent unauthorized access to the information in databases. Most importantly, the encryption technique works suitably, ensuring that the sender uses the public key for encryption and the receiver uses the private key to decrypt information. The keys must be handled separately, providing the information is further protected. The sender can share the public key without compromising the nature of the information since the key cannot be used to decrypt the information. However, the private key must be kept hidden at all times since it can open the information, revealing any underlying data.

The encryption model ensures that the pair of keys is first generated. The generation of the keys has to provide both public and private keys. Encryption is handled by converting the plaintext to ciphertext using the public key. The next step is ensuring that the information is handled well. The receiver can decrypt the information using the private key, creating a complete life cycle for the protected data.

Asymmetric encryption occurs in different models. There are different encryption methods that use the approach to ensure the safety and security of the information. Rivest–Shamir–Adleman (RSA) is a widely used method, enabling the safety of sensitive data and applying asymmetric algorithms to ensure long-standing information management [4]. The elliptic curve cryptography (ECC) works in a similar way to the RSA. However, it ensures shorter key lengths and a reduced storage requirement. ECC is also faster in computations, addressing the need to manage every marking scope and progressive information handling.

Asymmetric encryption can be applied to different models, ensuring suitable and appropriate community security development. In the first instance, the model applies to managing digital signatures, where they

CHAPTER 3 SECURITY PRINCIPLES

verify the integrity and authenticity of documents, messages, and software. Asymmetric encryption also ensures secure communications, with secure channels to safeguard browsing and other internet-enabled activities. Using SSL and TLS ensures the critical development of encryption to achieve the highest level of service as needed in every instance. Key exchange is also conducted using this approach, providing crucial advancements in targeting and marking safety for the communication protocol. This exchange mechanism appeals to having the best outcome in handling engagement and communication requirements.

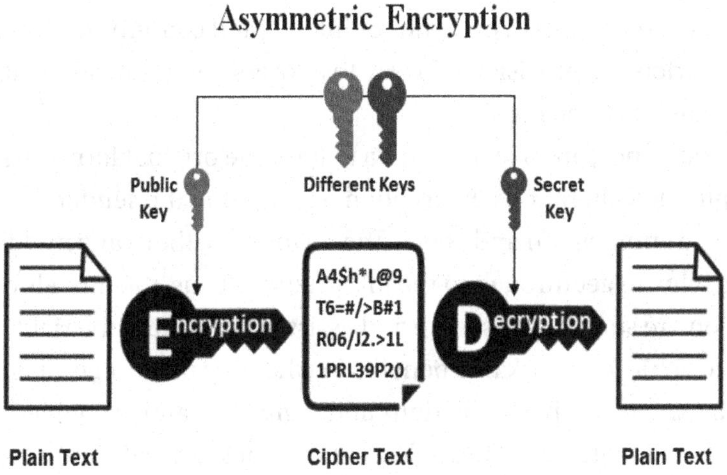

Asymmetric Encryption

Figure 3-2. Asymmetric Encryption Model of Action

Figure 3-2 indicates the model of action for asymmetric encryption where plaintext is converted to ciphertext using a public key. The figure illustrates the decryption of the ciphertext to plaintext using a secret/private key, helping to secure the information and communication process. This figure illustrates that the process ensures a critical engagement in having a similar pair of keys produced simultaneously and used to ensure sensible privacy in communication.

CHAPTER 3　SECURITY PRINCIPLES

Role of Encryption in Data Protection

Confidentiality: Encryption leads to a confidential transaction, where information is only accessible to the intended parties. This protects sensitive data such as personal information, confidential communication, and financial records. The process of encryption thus works to ensure that only authorized people have access to the desired information. Encryption while providing information ensures that even when security is breached, the attackers will lack the capacity and potential to understand and handle the provided communication. This indicates that data protection is assured even when data breaches or intercepted communication occurs since it restricts the provision of a way to address communication and engagement mechanisms.

Integrity: Encryption enables data to have the original format without any unintended alterations. Encryption ensures that the sender provides original communication and gets to the recipient without any middle parties having to get through to the information. Thus, the models of encryption are sustainably secure and assist in presenting the right values needed to achieve remarkable benefits in whatever mention is provided. Employing the use of hash functions and digital signatures enables the management of integrity. These elements can help address and addressee understand of providing communication to a level where they can conduct their services to achieve a remarkable outcome. The hash functions and digital signatures can be audited to help understand alterations to the communication within any provided instance.

Authentication: Encryption helps ensure that only the desired people can access the information. The process enables the development of a definition of services and activities meant to have private service and management at all engagement steps [5]. Using private and public keys prevents impersonation and creates a secure transaction platform. The private key ensures that only the recipient can decrypt the information and look into whatever they have received. The process enables a considerable

CHAPTER 3   SECURITY PRINCIPLES

confinement of the authorization to only the trusted party. Nonetheless, public keys must be provided with assigned keys, ensuring an individual has an ideal key to help them tabulate and present information to a desired level. The process makes it much easier to identify the recipient by providing certificates from trusted certificate authorities (CAS), which help to ensure that there is a level of safety aimed at ensuring the address for every information available to the users at any level.

Nonrepudiation: Using nonasymmetric encryption creates the option of ensuring digital signatures for information. This enables the traceability of information to confirm the sources, which makes it instrumental in legal transactions and financial activities, providing proof of engagement and communication between the parties. Digital signatures help by providing an audit trail that will help address emerging issues in the communication process management. Using digital signatures allows for the presentation of key communication channels, helping to account for the developing and engaging approach to handling communication between two parties. This helps with accountability and taking responsibility in key proportions that seek to enable a considerable adjustment of communication in every provided instance of communication.

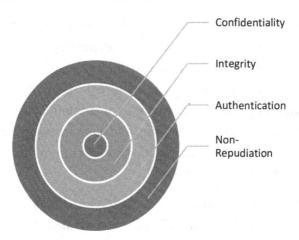

Figure 3-3. *Role of Values Linked Associated with Encryption*

Figure 3-3 indicates the different values of encryption in the organization. The figure illuminates their capacity to support the organization and exist alongside one another, leading to remarkable benefits in handling the organizational needs in managing data protection and security.

Authentication and Access Controls

Access controls and authentication are critical factors in addressing modern cybersecurity. Cybersecurity management is based on the capacity to protect systems, data storage, and handling by ensuring only authorized users can access the information. Addressing an organization or nation's resources works on the frame to securely handle the transmission of information to achieve the right outcome. Therefore, the application of this approach encourages users to use access control and authentication to safeguard their databases, ensuring confidentiality and integrity of data.

Authentication

This mechanism helps verify access to core information technology components of achieving the communication perioperative communication process, which works to mark engagement to handle authentication within the company to achieve the desired values [6]. In the first instance, three different categories of authentication will help any organization stand out in administering and working toward managing their access and handling of every platform they communicate on. The categories of authentication include:

- Something You Are: This authentication model uses biometric features like facial recognition or fingerprints. The model is personalized to individuals and helps them cater to different access levels.

- Something You Know: This enables the users to use their passwords and PINs, ensuring a more straightforward way to authorize an action and attend to various system requirements.

- Something You Have: This is entailed in physical devices that users must carry to help them identify and authenticate their presence; they include security tokens and smart cards that the individual parties use.

Password Policies

Passwords have to be administered with policies to enable their functionality and efficiency. Since passwords are the most common form of authentication, policies must allow their capacity to address safety and security, accounting for developing and handling every action toward administering the right password management system [7]. Different policies can apply to ensure the regulation of passwords in an organization, with a keen attempt to enable considerable development in addressing the safety and security of the systems. Some key password policies include:

- Length: This policy stipulates the number of characters appearing within the password to ensure safety and security. It is recommended that the password has a particular number on the system to enable its functionality. The number could vary from 4 to 12.

- Complexity: The password policy indicates the nature of characters within the password, helping to account for its purpose of addressing safety. Recommendations on upper- and lowercase letters alongside special characters and numbers dictate the capacity to achieve optimal safety with the password.

CHAPTER 3   SECURITY PRINCIPLES

- History: This password policy aims to enable critical safety in handling and addressing password management to prevent password recycling. This implies that users must use the policy that allows them to have passwords that have never been used on the system before, ensuring suitable safety capacities for the password.

- Expiration: This policy allows passwords to survive only for a depicted period. It indicates the right steps to take to understand passwords and prevent instances that might lead to compromised credentials being used on the system [8].

- Multifactor Authentication: This policy necessitates that users have an additional method of verification, which will help them address and move toward handling passwords and management. The passwords must be worked with to ensure achievement and address every scope of handling and managing the multiple verifications. An additional verification could include a one-time code sent to the user's email or phone number, helping stamp the system's security features.

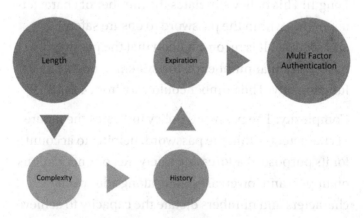

Figure 3-4. Password Policies

CHAPTER 3   SECURITY PRINCIPLES

Figure 3-4 indicates various strategies for creating passwords. The figure indicates the main factors for consideration, such as the length of the password, its complexity, installing expiration timelines for the password, the history of previously used passwords, and multifactor authentication for addressing the passwords. These factors ensure incremental management and handling of passwords in enterprises, ensuring suitable security measures for creating passwords. Thus, the figure highlights essential procedures in addressing and assessing passwords for their effectiveness upon creation in the organization.

Biometric Authentication

This authentication model uses a unique biological marker to help identify the users. This model ensures convenient use since they cannot be stolen and helps to provide a sustainable mechanism for addressing the safety of the systems, leading to much better engagements in using the system to achieve the desired level of safety and security. Notably, this approach enables the considerable development of a perspective that helps create the right element of value in administering safety and security through the capacity to address password management and generation. Key elements of biometric authentication include:

- Facial Recognition: This model ensures the integration of cameras and algorithms to analyze facial features, helping to authenticate users in the system.

- Fingerprints: This method uses scanners to capture fingerprint patterns. Every time a user signs in, the data is compared to the stored data, matching the prints and providing authentication for successful verifications.

57

- Voice Recognition: Algorithms analyze and store information about a person's vocal characteristics, enabling the data to be stored and compared with any attempt to sign in. This approach ensures accurate management of the effort to sign in, creating better verification means [9].
- Iris and Retina Scans: This model stores data on eye patterns, ensuring that every organization member is uniquely identified using these mechanisms. This helps to capture and relate to the development of safe authentication for users.

Biometric authentication is fast gaining relevance in modern times, given the difficulty of replicating biometric information. The possibility of high security has ensured that these approaches are used to provide definitive handling of the information given to help ensure authentication at all levels.

Access Control Models

Access control ensures that there is a designated level of addressing whoever has access to the system. This indicates that there are different levels of access to the system, implicating a person's level of access and capacity to be guaranteed an influence to handle information within the system. The levels of access include:

- **Discretionary Access Control (DAC):** This level of access is provided by the resource owner, who designs the system, outlining whoever has access to the system. This approach enables the right approach to dwell on critical individuals who can access the system. However, this approach has pertinent challenges in ensuring essential management of the underlying

CHAPTER 3 SECURITY PRINCIPLES

challenges in handling security permissions, as some people might need help understanding the scope of the permissions that they have been granted on the system. A significant issue with the access control model lies in the need for defined security management, making it prone to inconsistencies and errors, which affect the level of addressing safety and security for the different needs of the system. Scalability is also a key issue when using DAC. It requires intense management to ensure it can be applied across several platforms and networks, providing adjustment of every scope to achieve the most meaningful implementation. Thus, the use of DAC is limited to small organizations.

- **Mandatory Access Control (MAC):** This approach ensures critical policies emanate from a central authority. The access is granted based on the classification of databases within the system. Access also works by ensuring user clearances and addressing pertinent system modeling to achieve the designated value of attaining an influential outcome whenever needed [10]. The model is, therefore, safe for whoever has security clearance to handle information at a particular level, achieving a suitable engagement in remarking their development to become much notable in achieving the access mandates. MAC ensures uniform security throughout the organization, enabling considerable development in achieving better handling for access and authorization. This implies that it can be used by several users, making it suitable for both small and large organizations and promising the sustainable management of information provision at all levels. The use of MAC, however, is cumbersome

CHAPTER 3   SECURITY PRINCIPLES

to deploy in dynamic environments as organizations have to continually work toward addressing the rigidity in marking progressive handling of the MAC protocol. These challenges must be addressed during any implementation within large-scale organizations in a dynamic environment.

- **Role-Based Access Control (RBAC):** This access control mechanism ensures an accurate and modest step in determining incremental value in handling information at different levels. The RBAC enables considerable information management based on roles and duties that one carries out in the organization. The information is available based on groupings, implying that people within the same role in the company have the same level of clearance for information in the company. This is, therefore, concentrated on a generalized perspective, executed from the central position in the company, ensuring greater security and provision in managing the provision of information. The primary implementation needs for installing RBAC within the organization might be a task for the companies. However, the continued deployment within the organization makes it easy to use, as it replicates early instructions and helps define the scope of implementation of every valuable appeal at all levels. Using RBAC as a mechanism structure ensures accurate handling of every step demanded to achieve a sufficient model of addressing underlying needs within the company. Scalability for the model is also simple, as it uses the exact instructions and categories provided to establish the best protocol for handling employees and management in an organization.

- **Attribute-Based Access Control (ABAC):** Access control based on this approach is conducted based on various attributes. An organization outlines key attributes and measures that they have to work with, depicting and detailing individual conditions that have to be met to gain access to the network. The main attributes for selecting access include user attributes, environmental conditions, and resource attributes seeking to enable considerable handling of the control of when and why one has to access information on the system [11]. Thus, using ABAC considers several factors, all geared at ensuring the safety of the system information and achieving a reliable outcome in whatever conditions are necessary. ABAC is applicable in several environments because of the granularity of its implementation, enabling prolonged service management, which is based on the potential and capacity to ensure the administration of the best management levels alongside handling different requirements in the company. The approach is also vital in ensuring adaptation to organizational needs, remarking on the model's flexibility in achieving suitable advances in every selection needed. Therefore, the process marks a considerable advancement in pushing for safety and authentication measures in organizations. However, using the approach is complex as it requires several pieces of information to be keyed in before application in the company. This affects the nature of addressing organizational functions while implementing the development of the approach in the company.

CHAPTER 3   SECURITY PRINCIPLES

Implementation of Authentication and Access Control

Authentication and access control in the organization must be considered several times to ensure an instrumental implementation. Effective implementation has to work based on different practices that will ensure successful handling of the process to achieve a total capacity of serving and addressing various needs. The implementation process begins with assessment and planning, enabling the company to select its needs and critical resources on which it needs security. This is a crucial step to ensure that the authentication and access control are designed to enable sustainable development in critical proportions that help address significant system challenges. The selection of the suitable model depends on the size and demands of the organization, further leading to better handling and management of their requirements to achieve suitable engagement to value at all levels of appeal. The growing attention and modeling of the specific elements of the organization lies within the decision point that helps to develop the right policies that must be followed to ensure refined management of authentication and access control. Organizations also have to follow critical practices to ensure successful implementation. These practices are key to designating the development and growth of the company in essential mechanisms that assist in catering to the depiction of valuable advancement for the company to ensure implementation of the right access and authentication protocol to help safely handle its information. The essential practices to ensure a successful implementation include:

- **Regular Audits:** Organizations must consistently carry out audits, giving them the proper control over the different functionalities and critical roles in the organization. This approach ensures that they can adjust permissions and enhance the company's role provisions even when an employee has been fired.

CHAPTER 3   SECURITY PRINCIPLES

- **Monitoring and Logging:** Monitoring is a key step that enhances the achievement of model appeal in targeting and working on activities within the system. Monitoring makes it much easier to look into access attempts and activities to ensure any unauthorized access and suspicious behavior are handled within the right level of achieving suitable management for the organizational network.

- **User Training**: This approach ensures that users from the management and employee segments receive increasingly beneficial information on security features. It grants them the chance to develop proactive safety habits, helps them recognize phishing attempts, has strong passwords, and safeguards their system to a required level [12].

- **Least Privilege Principle:** This principle ensures that a user is granted a minimum level of access, helping them to carry out their activities on the company network system easily.

- **Access Reviews**: A policy to review access regularly will ensure the revision of access roles, execution of network activities, and revocation of network privileges whenever needed. This policy is key to addressing the changing phase of organizational security, granting a much-needed approach to sustaining safety in the company.

Figure 3-5. *Implementation of Access Control*

Figure 3-5 indicates the different processes used when implementing access control, such as regular audits, monitoring, user training, the least privilege principle, and access reviews. These processes help provide remarkable management and handling of the access control mechanisms, offering a better understanding of the entire scope of access control regulations. Each of these mechanisms ensures stellar access control modeling, providing suitable adjustments to the access control process.

Integrity Checks and Other Security Measures

In the modern digital age, data integrity and security are key factors that organizations strive to achieve a sustainable outcome at all times. Notably, managing integrity checks and security measures encourages data handling in key models, aiding the platform's generation and management of safety. Different models have been applied over the ages to help address the nature of safety and security modeling in organizations. They are advancing to a level where different organizational advances are crafted to achieve the required outcome. Different mechanisms can be applied to ensure more security and safety within the organizations.

CHAPTER 3   SECURITY PRINCIPLES

- **Hashing:** Hashing involves using a hash code to help address verification. The process of hashing is an instrumental step in helping with handling verification. A hash code is compared against the original, enabling instrumental analysis of the hash code and ways it can be verified. Hashing ensures that the data has not been tampered with or altered, leading to an insight into the handling and managing of the information on the platform [13]. Different hashing algorithms are applied, with crucial insight into the designated outcome in addressing security. Key algorithms used include SHA-1, MD5, SHA-2, and SHA-256. The mechanism applies to different platforms, such as password storage, blockchain technology integrity, and file integrity verification.

- **Digital Signatures:** Digital signatures help verify the authenticity of documents and digital messages. The process uses public-key cryptography involving both public and private keys. The recipients use the private key to decrypt information, ensuring that they can understand the integrity and authenticity of the document or communication. Digital signatures often carry legal validity, provided one can use them to sign documents electronically and successfully handle financial and legal documents.

- **Firewalls** are network security systems used to monitor and control traffic based on predetermined regulations. Firewalls deal with incoming and outgoing traffic, adding to the deployment of safety measures within the network. The firewall acts as a way to foster the safety of the internal network against unsecured external networks. Thus, they serve to prevent unauthorized access and cyberattacks. Different types of firewalls are used to enhance security, depending on a defined target and engagement appeal. Significant firewalls include packet-filtering firewalls, proxy firewalls, stateful inspection firewalls, and next-generation firewalls. They can be used from individual and organizational computers to enable security and safe engagement within these networks.

- **Antivirus Software:** This software detects, removes, and prevents any instance of malware, trojans, and worms. They enhance the signature-based detection used in identifying known malware [14]. They also have the best scope and capacity to address heuristic-based detection to analyze programs and see whatever new programs have malware within them. However, antivirus software demands regular updates that will help them identify new threats and provide suitable patches to prevent any penetration of viruses.

- **Intrusion Detection and Prevention Systems:** These are systems installed within the network to help monitor anomalies, identify suspicious activity, and handle potential threats. The use of the approach enables critical identities to help monitor, identify, and alert the system on any ongoing attacks. The approach also works to suitably help in marking preventive measures that are brought in to help achieve safety modeling and engagement. Therefore, the right approach in managing and working with the IDS brings out real-time monitoring, response, and addressing critical issues affecting financial institutions and government agencies.

CHAPTER 3   SECURITY PRINCIPLES

- **Security Information and Event Management (SIEM):** This system helps address real-time analysis of security alerts from network hardware and applications. SIEM has tools that collect, assess, and compare security data from all devices within the network, bringing critical assistance in detailing anomalies within the system and marking potential threats to the network. The use of SIEM helps to structure reports about security issues within the network, helping to bring about better analysis and assessment of the system. This approach makes the auditing process easier to conduct. It offers a multistep approach to depict an increasingly beneficial way to attain valuable advancement in managing security issues within the network. Therefore, SIEM works by enabling companies to achieve situational awareness and work on crucial incident responses that can assist them in learning about the suitable activities to be conducted at any given point.

- **Data Loss Prevention:** This is a collection of strategies and designated tools that work together to prevent unauthorized transmission, access, and use of sensitive information in the company. The approach works to ensure that they monitor data movement across the network, leading to the prevention of potential data breaches and leaks at any given point. DLP defines sensitive data within the network, helping enforce policies and regulations to monitor sensitive data and prevent malicious activities on the platform.

CHAPTER 3　SECURITY PRINCIPLES

- **Regular Security Audits and Vulnerability Assessment:** This security approach aims to identify and mitigate the security system's weaknesses. The approach ensures a comprehensive analysis of the organization's policies, processes, and controls to adequately handle sensitive information and data. The approach also leads to critical handling of vulnerabilities, with the main aim of measuring the organization's required standard, each mentioning helping to safeguard the system against being used by attackers [15]. The approach is crucial in designing and enabling critical adjustments to handle automated tools that scan for vulnerabilities. The assessment also includes simulations on attacks and potential breaches, seeing the channel of presenting and administering valuable engagement to consider greater growth in addressing the pertinent issues. Therefore, the approach marks a valuable growth of strategies and procedures that assess the security system's strength and assess for vulnerabilities within the system.

- **Patch Management:** This approach ensures that updates to the system are conducted on time to help fix known vulnerabilities and weak points within the network system. Patch management offers a chance to ensure timely updates, assess the patches before deployment, and achieve updated system modeling to address consistent management of the needed adjustments at any level of engaging their demands. Patch management is, therefore, key to ensuring that the organization has up-to-date systems that assist in creating the most remarkable system, helping it administer value across all segments and achieve valuable system functionalities.

- **Endpoint Protection:** This practice encompasses strategies to secure endpoints such as laptops, tablets, smartphones, and IoT devices connecting to the organizational network. Different tools can be applied to ensure protection, leveraging the value of information and assisting in crafting the right outcome. Tools such as endpoint detection systems (EDR), antivirus software, and antimalware software work within the right framework to address weak points and enable critical developments to achieve suitable results as demanded.

- **Network Segmentation:** This practice helps protect the organization's excellent network. The approach encompasses smaller segments within the network, each with its security controls. The main objective is to spread malware and unauthorized access within the network system. The implementation model is virtual local area networks and subnetting. Organizations can ensure the achievement of the right outcome by isolating critical systems and information from being used alongside regular company departments.

Even with these applications and approaches, companies demand to have the right model of regulation and control, spanning a combination of these efforts and fundamental approaches such as digital signatures and hashing. Companies must create robust and multilayered defense systems that increase their resilience against attacks and ensure overall security in the organization. The mechanism creates a relevant step to assist in handling threats, achieving confidentiality, and handling the availability of critical systems as demanded by the organization. Thus, using these models assists in creating the best identification and management of threats within the organizational networks.

Summary

Encryption is an essential element in modern society; using encryption ensures the definitive management and address of various security needs in companies. Modern companies use different encryption models. Symmetrical encryption involves using a public key, enabling users to decrypt information sent over the networks. Famous methods of symmetrical encryption include AES, DES, and 3DES. Asymmetrical encryption involves using a public and private key, implying the communication recipient has to use their private key to decrypt the information. More to the point, encryption has to be safeguarded since public keys in symmetrical encryption can potentially result in unauthorized information access.

Moreover, authentication and access control are approaches that are applied to help in managing network systems. Authentication can be applied based on passwords and biometric models that help verify the systems' various users. On the one hand, access control uses avenues such as role-based and attribute-based models. Role-based models ensure a definition and handling of the responsibilities of every member of the organization within the network, ensuring more excellent attendance for privileges on the platform. Attribute-based models evaluate features such as environmental conditions, user attributes, and resource attributes; each approach enables functional development to achieve critical outcomes in handling the company's security.

Further, this chapter indicates additional security approaches that can be used to engage networks and companies, ensuring appropriate outcomes in each instance; these additional approaches include patch management, network segmentation, data loss prevention, intrusion detection systems, and endpoint security. Integrating the features of these models guarantees greater security to the network, enabling more significant success in whatever activities they partake in. Thus, security principles must be addressed to ensure that organizations always have the best protection and data safety approaches.

References

[1] P. Dixit, A. K. Gupta, M. C. Trivedi, and V. K. Yadav, "Traditional and hybrid encryption techniques: a survey," in *Networking Communication and Data Knowledge Engineering: Volume 2*, Singapore: Springer, 2018, pp. 239–248.

[2] W. Easttom, *Modern cryptography: applied mathematics for Encryption and information security*. Springer Nature, 2022.

[3] M. N. Alenezi, H. Alabdulrazzaq, and N. Q. Mohammad, "Symmetric encryption algorithms: Review and evaluation study," *International Journal of Communication Networks and Information Security*, vol. 12, no. 2, pp. 256–272, 2020.

[4] I. Meraouche, S. Dutta, H. Tan, and K. Sakurai, "Learning asymmetric encryption using adversarial neural networks," *Engineering Applications of Artificial Intelligence*, vol. 123, p. 106220, 2023.

[5] N. Mohammad, "Encryption Strategies for Protecting Data in SaaS Applications," *Journal of Computer Engineering and Technology (JCET)*, vol. 5, no. 1, 2022.

[6] W. He et al., "Rethinking Access Control and Authentication for the Home Internet of Things (IoT)," in *27th USENIX Security Symposium (USENIX Security 18)*, 2018, pp. 255–272.

[7] E. Stobert and R. Biddle, "The password life cycle," *ACM Transactions on Privacy and Security (TOPS)*, vol. 21, no. 3, pp. 1–32, 2018.

[8] M. Yıldırım and I. Mackie, "Encouraging users to improve password security and memorability," *International Journal of Information Security*, vol. 18, pp. 741–759, 2019.

[9] Z. Rui and Z. Yan, "A survey on biometric authentication: Toward secure and privacy-preserving identification," *IEEE Access*, vol. 7, pp. 5994–6009, 2018.

[10] A. Alshehri and R. Sandhu, "Access control models for cloud-enabled internet of things: A proposed architecture and research agenda," in *2016 IEEE 2nd International Conference on Collaboration and Internet Computing (CIC)*, 2016, pp. 530–538.

[11] A. Bierman and M. Bjorklund, "Network Configuration Access Control Model," no. RFC 8341, 2018.

[12] S. Behrad, E. Bertin, S. Tuffin, and N. Crespi, "A new scalable authentication and access control mechanism for 5G-based IoT," *Future Generation Computer Systems*, vol. 108, pp. 46–61, 2020.

[13] X. Luo et al., "A survey on deep hashing methods," *ACM Transactions on Knowledge Discovery from Data*, vol. 17, no. 1, pp. 1–50, 2023.

[14] F. A. Garba et al., "Evaluating the state of the art antivirus evasion tools on Windows and Android platform," in *2019 2nd International Conference of the IEEE Nigeria Computer Chapter (NigeriaComputConf)*, 2019, pp. 1–4.

[15] S. Bozkus Kahyaoglu and K. Caliyurt, "Cyber security assurance process from the internal audit perspective," *Managerial Auditing Journal*, vol. 33, no. 4, pp. 360–376, 2018.

CHAPTER 4

Authentication in Cybersecurity

Author:

Saurav Bhattacharya

Introduction

Authentication is the process of validating the identity of a user by usually obtaining some sort of credentials and making sure that they match the user's provided login information, in order to produce some type of output which shows some sort of validation that the user is who they say they are. Quite simply, authentication is about making sure that a message, transmission, or action is actually coming from the person or entity that it is claimed to be coming from. When it comes to an environment such as the internet, implementing it where we have various users and all of them will have different types of resources that they want to access, it becomes of the utmost importance since it ensures that only authorized persons can access the resources. With the boom in technology and the massive development of the internet, we have seen more and more resources moving from a localized type of use to more of a web-based type of use, and this leads to an increased need for stronger and more multileveled methods of authentication in order to protect those resources. Another reason why authentication is important in today's world is the sheer number of attacks and attempted security breaches that exist. Nowadays, the two most widely used authentication techniques are username and password, and challenge and response. These techniques are used to validate one's identity and grant access to the correct resources if the credentials given are valid. However, it is recognized that in order for one to implement stronger methods of authentication, a third party will be required in order to verify the user's identity or personal information. Multifactor authentication (also known as two-factor or dual-factor authentication) can be defined as a method of confirmation of a user's claimed identity in which a computer user is granted access only after successfully presenting two or more pieces of evidence to an authentication mechanism: knowledge, possession, and inherence. Two-step verification is a technique designed to achieve a goal similar to multifactor authentication by means of a method of confirmation that involves two different methods of verifying the user's claimed identity. Any combination of methods can be used as related to two-step verification, including the more traditional username and password, a text message sent via mobile phone, a secondary password provided by a security token or smartcard, or any other first or second-level methods that can be considered.

Definition of Authentication

Authentication is the process by which the system confirms that the user attempting to access a resource is actually who they claim to be. In simple terms, authentication is the process of verifying the identity of a user. When a user logs into a computer, a website, or an application, or when they try to access an email account or make an online transaction, they need to prove that they are the person that they say they are. This is essentially the main purpose of authentication. In many network and web-based systems, the purpose of authentication is to provide proof of identity. The username or user ID is a unique identifier and so belongs to the category of "something you know." User IDs are often widely known, meaning that protecting them with a password is very important. A password belongs to the category of "something you know" and so does a PIN number.

CHAPTER 4   AUTHENTICATION IN CYBERSECURITY

Together, a user ID combined with a password or PIN number may help to provide stronger verification of a person's claimed identity. Nowadays, public key cryptographic techniques are increasingly being used to provide a very high level of reliable verification, without inconveniencing users themselves (Karim et al. 2023). The use of digital signatures and digital certificates has helped to create the possibility of using cryptographic verification techniques within widely available internet-based applications and resources. Public key encryption involves two keys, a public key and a private key. The public key can be shared; the private key should not. A digital signature uses the sender's private key to create a unique string of characters that act as a digital stamp of authentication. This string can be verified by using the sender's public key. As well as being used to provide verification in digital signatures, public key cryptography can also be used to provide additional security in other authentication methods. The use of biometric methods, such as fingerprint or eye recognition technology, can provide very strong verification of a user's identity. Such methods can also provide a straightforward and convenient means for the user to prove their claimed identity without having to remember numerous complicated passwords and user IDs. The use of biometrics involves confirming measurements taken from the user against records held in a database. Biometric methods are increasingly being used in high-security locations, such as immigration checkpoints, building entrances, and airports.

Importance of Authentication in Cybersecurity

Maintaining a good understanding about the importance of authentication in the context of cybersecurity and keeping updated about emerging authentication technologies and threats are essential for cybersecurity professionals. Well-informed security decisions and strategies are vital for organizations to protect against the exponentially growing landscape of cyberattacks. The next section will delve into different methods and processes by which a user's identity can be confirmed through authentication.

CHAPTER 4   AUTHENTICATION IN CYBERSECURITY

Over the years, a large number of security incidents and data breaches have involved the exploitation of weak authentication. For example, a report from Verizon indicates that over 80% of hacking-related breaches involve the use of lost, stolen, or weak passwords (Dastane 2020). Inadequate authentication can leave unauthorized individuals with the opportunity to access systems and sensitive information and seriously impact the confidentiality, integrity, and availability of data. Ensuring effective cybersecurity, therefore, heavily relies on the application of best practices in authentication. By knowing the significance of authentication and employing best practices in a variety of forms, including implementing multifactor authentication and using complex passwords, organizations can better protect their systems and data from cyber threats.

An effective cybersecurity method comprises multiple layers of protection dispersed across the computers, networks, and programs. This approach is very useful because each layer can catch different types of threats and together they create a powerful defense against cyberattacks. Authentication is the essential element for each layer of protection. It is the process of verifying the identity of a person or a computer that is attempting to access a system. Effective authentication assures that an entity is who it claims to be and ensures that the entity has the proper authorization to access the system. For cybersecurity, the importance of authentication cannot be overemphasized. It is the cornerstone for securing computer systems and networks across the globe.

Types of Authentication

There are multiple types of authentication, each with its own benefits and drawbacks. The most basic form of authentication is password-based authentication, where users are required to prove their identity by providing the correct password. While this method is simple and easy to implement, it is often considered to be less secure compared to other methods, especially if the user has a weak password. Two-factor authentication, on the other hand, is a type of multistep authentication, which requires two different forms of identification to prove the identity of a user. The most common factors used with two-factor authentication are "something you know" (like a password or PIN) and "something you have" (like a smart card or token). By requiring two different forms of verification from a user, this method can significantly increase the level of security. Biometric authentication is a security process that relies on the unique biological characteristics of an individual to verify that he is who he says he is. There are many different types of biometric systems, including fingerprint readers, retinal scanners, and voice recognition systems. These technologies are becoming increasingly popular, especially in mobile devices where integrated fingerprint readers are commonly used. Multifactor authentication is an authentication method in which a computer user is granted access only after successfully presenting two or more pieces of evidence to an authentication mechanism: knowledge, possession, and inherence. Multifactor authentication is a critical component when it comes to preventing unauthorized access to sensitive information and financial transactions. By boosting the security of user logins and data access, multifactor authentication can help organizations comply with regulatory requirements like GDPR and HIPAA (Thapa & Camtepe 2021). When choosing an authentication method or combination of methods, it is important to consider various factors, including the level of security required, the value of the resources being protected, and the cost and practicality of implementation. This is particularly true in the case of businesses and large organizations, where various factors must be balanced and considered before any changes or investments are made.

Password-Based Authentication

This is the most common type of authentication method used. Users are required to input their username and password in order to gain access. After entering the correct details, the user is successfully authenticated and allowed entry into the system. Usernames and passwords must be kept secret and should not be revealed to anyone. Passwords are the most common method of authentication because they are more convenient and can be used in remote access. However, most password-based authentications are vulnerable to some type of attacks like brute force attacks, dictionary attacks, and keylogger attacks. Additionally, if a user's password is ever discovered or stolen, such as through social engineering or phishing attacks, an adversary will be able to gain access to the system just as if they knew the correct password all along. Lastly, it is necessary for a system to store a user's password to allow for future verification as they log in. To maintain security, the actual passwords are not stored - instead, a "password hash" is generated from the password and it is this value that is kept. However, if an adversary is able to access the password hashes for a system, they can attempt to use password attacks to retrieve the original passwords, which would then compromise the security of all the user accounts on that system. The most significant vulnerability of password-based authentication is that it relies on users choosing and not revealing a strong and unique password for their account. If a user's password is weak or easily guessable, it becomes much easier for an attacker to gain unauthorized access to the system. There are many different types of password attacks that involve an adversary trying to find out a user's password. For example, in a "brute force" attack, an adversary will systematically try every possible password combination until the correct one is found. These attacks can be made more efficient by the use of "rainbow tables," which are precomputed tables of password hashes that can be used to look up an easily reversible hash and find a corresponding password. Another approach is to launch a "dictionary" attack, in which the adversary tries a list of common words and passwords in the hope that a user has chosen one of these. These sorts of attacks can be used to recover user accounts *en masse* if a security compromise results in the theft of password hashes from a system.

Two-Factor Authentication

Two-factor authentication is an increasingly common practice these days and is often used as an additional layer of security. As its name suggests, it requires two different forms of identification. The first factor is something you know, like how a normal password would be used. The second is something you have, which could be anything from a mobile device to a bank card. This means that anyone trying to hack into an account would not only have to obtain the password itself, but they would also have to gain access to the second form of identification at the same time. As a result, two-factor authentication is a great way to add an extra level of security to your accounts. In fact, using it can make your account up to 99% less likely to be hacked into. It has been widely recommended by data security experts for a long time. The basic principle of two-factor authentication has been around for decades. In the past, security experts would have referred to it using all sorts of different names and terms, such as "multilevel authentication" or "2FA" for short. However, the concept of using two steps to verify the legitimacy of a user has become increasingly modernized and is now widely known and implemented by companies and organizations in all different kinds of sectors around the world. Although the use of two-factor authentication can't prevent all forms of hacking or cyberattacks, that significant reduction in risk that it offers makes it an essential consideration for anyone looking to improve their data security. Even if you are not required to use two-factor authentication for certain accounts, such as your emails or social media, it is definitely still worth going through the process of setting it up. It only takes a few minutes and is a small inconvenience when compared to the potential fallout from a major data breach. By showing that you are aware of this important security measure and taking active steps to protect your own data, you'll also be helping to promote a wider understanding and acceptance of two-factor authentication.

CHAPTER 4   AUTHENTICATION IN CYBERSECURITY

Biometric Authentication

Biometric authentication relies on the unique biological characteristics of an individual to verify that the person is who they claim to be. Different physical or behavioral biometric factors can be utilized for authentication, such as fingerprints, facial features, or iris patterns. The most common biometric authentication method is fingerprint scanning. A fingerprint scanner takes an image of a person's fingerprint and analyzes the unique patterns present, such as arches, loops, and whorls. When a user wants to authenticate, he or she will place their finger on the scanner, which will take an image and create a template by identifying distinctive features. This template is then compared to the stored template associated with the user. If the two templates successfully match, the user will be granted access. However, it is important to note that some weaknesses exist with biometric authentication. First, although biometric data cannot be easily stolen and used in the same way that a username and password can, if it is compromised, it is almost impossible to replace. For example, if a person's fingerprint is replicated and used to unlock a device, there is no way for that person to change his or her fingerprint, leading to a severe privacy and security issue. Second, biometric authentication devices have a limited number of templates that they can store. This means that these devices may be prone to attacks that involve providing a large number of different biometric samples in an attempt to match a stored template. This is known as a "spoofing attack" and, to prevent it, the device needs to be sophisticated enough to detect the difference between a live biometric sample and a fake one. In the case of fingerprint recognition, this may involve measuring the capacitance between the ridges and the valleys of a fingerprint to detect an electrically conductive pattern (Maltoni et al. 2022). Third, a biometric authentication system should always be used in conjunction with other security measures, such as encryption and secure digital storage, to ensure that the biometric data cannot be easily stolen and exploited. Biometric authentication is widely used in various sectors, from forensic analysis in law enforcement to security access in building control. With the advancement of technology, biometric methods will continue to develop, making it quicker and easier to accurately confirm the identity of an individual.

Multifactor Authentication

Multifactor authentication (MFA) is a method of authentication that requires more than one verification method and adds a critical second layer of security to user sign-ins and transactions. This is achieved by requiring two or more of the following authentication methods: something you know, something you have, and something you are. For example, when you use an ATM, you need both your bank card (something you have) and your PIN (something you know). In another common real-world example, if a door is protected by a MFA security system, it could require both a keycard (something you have) and a fingerprint scan (something you are) to successfully grant access to the holder. In the context of digital systems, the most common MFA methods include a combination of a password and a code sent to a mobile phone, a password and a fingerprint scan, or a password and a smart card that is inserted into the computer. By introducing MFA, even if an attacker has compromised one of the authentication methods, they will still be blocked from accessing the system because the additional methods remain secure. According to research from Microsoft, MFA can block over 99.9% of account compromise attacks (Rains 2020). In addition to protecting against password attacks, MFA is also effective in mitigating other common authentication vulnerabilities. For example, if an attacker attempts to use a phishing website to steal a user's password, the stolen credentials would be useless without also having the additional MFA method. Overall, the growing prevalence of cyberattacks and the vulnerabilities of single-factor authentication methods have made MFA an increasingly important consideration in ensuring the security of digital systems. It is advised to deploy MFA for all user accounts, especially those with administrative rights and those storing sensitive or regulated data. However, it is recognized that implementing MFA can require additional time, money, and effort for an organization. As a result, the decision to introduce and enforce multifactor authentication may need to balance the critical need for security with any potential limitations. This is likely to involve a thorough risk assessment and a consideration of relevant industry standards and regulations.

Common Authentication Vulnerabilities

In the vast realm of cybersecurity, there exist various common authentication vulnerabilities. These weaknesses pose grave risks to the security of our digital world. By exploiting these vulnerabilities, malicious actors can gain unauthorized access to sensitive information or even take control of an entire system. Understanding and addressing these vulnerabilities is crucial in safeguarding our online systems and data. In this section, we will explore some of the key authentication vulnerabilities that commonly arise and discuss ways to mitigate them.

Password Weaknesses

Passwords are a common and vulnerable security loophole, making them an easy target for unauthorized access. To enhance security, users should choose strong and regularly updated passwords. It is crucial to educate users to report any password compromises. Brute force and dictionary attacks are frequently used by attackers to crack passwords. Adding "salting" to passwords significantly improves their protection by making them harder to break (Rathod et al. 2020). Using unique salts for each user prevents attackers from cracking multiple passwords with precomputed hash tables.

Credential Theft

It is important to understand the risk of having our credentials stolen. Unauthorized access to sensitive data and systems can be gained from stealing user credentials. The most common way is stealing usernames and passwords. Attackers can obtain credentials by installing spyware or logging software on a user's computer or by sniffing unencrypted traffic on the data network. Once stolen, these pilfered credentials can be widely used by the attacker to move laterally within the compromised organization. In many cases, attackers are able to gain initial access to a system and steal credentials using valid accounts and existing remote accesses. For example, insiders at an organization are often the culprits for stealing credentials to aid the attacker. A study of data breaches shows that 34% of breaches involved internal actors. Another example is hacking groups including advanced persistent threat (APT) actors. These groups have been known to conduct sophisticated attacks while maintaining a presence on a network for a long period of time by employing stealthy tactics. One of the main tactics used by these hacking groups is stealing or capturing credentials through remote access trojans and keyloggers. It is often the case where security measures such as firewalls and intrusion detection systems cannot prevent attacks or detect the presence of these remote access trojans due to the legit nature of the communication channel through the use of stolen credentials. Well-defined security policies and ongoing security measures can help prevent and detect credential theft activities. Close monitoring of remote access logs and regular security education and awareness to system users are effective ways to detect and prevent credential theft. We will discuss more on this in the last section on best practices for secure authentication.

Social Engineering Attacks

Unfortunately, no matter how strong a password might be, it is often the case that the users themselves are targeted and the weakness lies in the user rather than the password. This is where social engineering attacks come into play; using deception to manipulate individuals into divulging confidential or personal information that may be used for fraudulent purposes. This is a well-established technique and has been used in many high-profile cyberattacks, including the hacking of Twitter accounts for Barack Obama and Joe Biden in July 2020 (Witman & Mackelprang 2022). There are many forms of social engineering, but the main kind typically seen in a cybersecurity context can be split into three different types. The first is "human-based," where trust and real personal connections are used to get information. This can be a long game, building up trust over time, or it can be a much quicker attack, for example, by just walking into and pretending to be a trusted person in a building. The second is "computer-based," which uses ways to trick the user into doing things on a computer which they should not, such as clicking on a bad link. And third, there is "blagging" or "pretexting," where someone directly or indirectly poses as a legitimate person and asks for information. This is something common in more elaborate attacks, and often it is used to gain information in order to then further trick a user or access accounts. Well-informed user education and awareness, therefore, is one of the best defenses against social engineering attacks.

Brute Force Attacks

A brute force attack can manifest itself in many different ways, but the most well-known instance is when a hacker uses a software program or script to submit a large number of guesses in rapid succession. In the context of a web application or a password-protected area, the attack will try username and password combinations until it gets a match. This is why

CHAPTER 4   AUTHENTICATION IN CYBERSECURITY

a strong password policy is incredibly important for thwarting brute force attacks – the longer and more complex the password, the harder it will be for anyone to crack it using a brute force attack. Implementing an account lockout policy can also help to reduce the risk of a brute force attack being successful. For example, an account lockout policy might dictate that, after a certain number of incorrect login attempts, a user account is locked and cannot be restored without the intervention of an administrator. For web applications, an account lockout policy combined with measures such as CAPTCHA or a two-factor authentication requirement can substantially reduce the risk of a brute force attack. Although a brute force attack is often illustrated in the context of an attacker trying to guess a password, the possibility of a similar attack on other forms of authentication should not be overlooked. For example, an attacker could try every possible encryption key in existence using a brute force attack against an encrypted block of data. Deciding on an appropriate security response to a brute force attack, or successfully avoiding one in the first place, largely depends on the specifics of the attacker's tactics and the nature of the security system in question. For example, when considering how to build a security system in such a way as to make it resistant against a brute force attack, it is best to seek advice from a qualified cybersecurity professional.

Best Practices for Secure Authentication

The section on best practices for secure authentication aims to provide valuable insights and recommendations. In this section, we will explore various strategies to enhance the security of authentication processes. These practices will help organizations effectively protect sensitive data and prevent unauthorized access. By implementing strong and robust authentication methods, businesses can mitigate the risks of potential security breaches. It is crucial for organizations to adopt these best practices to ensure a secure and reliable authentication system.

CHAPTER 4   AUTHENTICATION IN CYBERSECURITY

Strong Password Policies

Each method of authentication has its own vulnerabilities. For instance, a weak password has a much lower direct cost in terms of how long a brute force attack would take to be successful than other methods such as a smart card being lost or stolen. Also, certain types of authentication are better suited to certain requirements. For example, a system which constantly needs to verify and reverify a user's identity hundreds of times a second, such as a scenario within a busy IT network, suitable biometric recognition may be impractical due to both the time taken to scan and verify a user's identity and also the potential initial cost and maintenance of the associated scanning hardware.

Users should be educated on the importance of using strong passwords and how to create and maintain them. For example, a strong password should ideally contain twelve or more characters, be a passphrase that is easy for the user to remember but hard for others to guess, and not contain any word found in a dictionary in any language. Employing strong password policies is crucial as passwords are a popular target for cybercriminals trying to gain unauthorized access to systems. Organizations should enforce the following password policy requirements to ensure the effectiveness of user passwords:

- Preventing the use of the same password: There should be a policy to disallow the reuse of previous passwords for a certain number of generations to reduce the likelihood of compromised passwords being repeatedly used.

- Avoiding the use of easily guessable information: Users should be advised against using passwords based on information that can be easily guessed, such as "password," "123456," "qwerty," "abc123," and common words or phrases like "admin," "login," "football," "iloveyou," "letmein," etc.

- Password complexity requirements: To enhance the strength of passwords, it is advisable to enforce the inclusion of characters from at least three of the following categories: uppercase letters (A-Z), lowercase letters (a-z), numbers (0-9), and special characters (e.g., ! @ # $ %).

- Minimum password length: Passwords should be a minimum of eight to ten characters long as longer passwords are generally more secure.

Regular Password Updates

Regularly updating passwords decreases the likelihood of successful password attacks. For example, even if someone got a hold of an old password, if it's been changed since, that old password would no longer be useful. It is common to see advice and systems that enforce regular password updates. For example, some organizations require their staff to change their passwords every 45-60 days. After a password has been set for a user, many systems and applications will keep track of when it was last changed. When that user next logs in using that password, the system will check how long it has been since that password was last updated. If it has been a certain length of time since the password was last changed, the user will be prompted to choose a new password before they can continue. This helps ensure that the user creates a fresh password that could increase the security of their account. Over time, however, the passwords that users choose could become less secure. For example, users might choose passwords that are slight variations of their old ones, or they might choose new passwords that are easily guessable because such passwords are easier to remember. Another issue is that studies have suggested that forcing users to adopt new passwords regularly might perversely encourage practices that decrease password security, such as writing them down on Post-it notes, sharing them, or always basing passwords on a simple, memorable root with just a numeric suffix that changes each time. If users are repeatedly forced to change their passwords, these practices can become very hard for a system administrator to police. As such, it's essential to strike a balance with regular password updates: setting the time period too short might not provide a meaningful improvement to security and could encourage bad habits, like choosing very simple, similar passwords. On the other hand, setting the period too long could mean that compromised passwords aren't changed for a significant amount of time after an attacker has used them, and the protection that regular updates provide is lost. Periodic reviews of the password policy are necessary to consider the keystroke dynamics of the password, aging, a review of the policy to keep pace with changing technology and user behaviors, and the least resistance to attacks. So, regular password updates can provide a useful defense mechanism against unauthorized users.

User Education and Awareness

Security education and awareness play a pivotal role in helping individuals understand the importance of authentication and how to protect their credentials. Providing interactive cybersecurity exercises and organizing cybersecurity themed occasions, for example, security awareness week, are effective ways to help raise public awareness of cybersecurity. Security awareness training for employees is also critical in helping to develop a culture of security within an organization. A comprehensive training program should integrate ongoing awareness training, for example, monthly updates, into its strategy and should cover aspects, for example, how to report suspected phishing emails. In the meantime, employers should use a mixture of learning tools and resources as using only PowerPoint presentations is likely to be forgetful and considered boring to employees. A training program's effectiveness should also be evaluated consistently and techniques, like phishing simulations, can be used as a metric to measure the improvements in user awareness. By regularly sending fake phishing emails to employees, a company will be able to track how many people open the email and proceed to click the malicious link. Over time, as the training program takes effect and employees become more aware, this number should reduce. This will also furnish an employer with a constructive indication of whether the training requires updating and improving. By attending and participating in security training, both employees will demonstrate a commitment to self-improvement and their employer will be able to demonstrate to clients and customers that appropriate training and measures are in place to protect data.

Implementing Multifactor Authentication

Implementing multifactor authentication significantly enhances security by requiring multiple forms of authentication before granting access. Users must successfully present two or more separate pieces of evidence to verify their identity. This approach effectively strengthens the security posture by adding an extra layer of defense against unauthorized access. Even if an attacker manages to steal a user's credentials, such as a password or a PIN, it is not sufficient to pass the multifactor authentication check.

Multifactor authentication can be implemented in various ways. For example, after entering a password, the system may prompt the user to enter a unique code sent to their mobile device. In another scenario, the user may be required to submit a fingerprint in addition to entering a PIN. The multiple authentication factors used in multifactor authentication can be categorized into three types: knowledge factors, possession factors, and inherence factors. Knowledge factors refer to something the user knows, such as a password, a PIN, or the answer to a security question. Possession factors relate to something the user has, like a smart card, a token generator, or a mobile phone. Inherence factors are biometric properties tied to the user, such as fingerprints, facial features, or iris patterns.

In addition to user accounts, multifactor authentication can also be applied to system-level access, such as remote access to servers or network devices. Moreover, with the increasing adoption of cloud services, many cloud service providers offer built-in support for multifactor authentication as an additional security option for customer accounts. Overall, multifactor authentication is an effective and practical method to enhance security that has gained widespread acceptance across different industries and sectors. Its ability to provide an additional layer of defense to protect against the increasing number of security breaches resulted from stolen or weak authentication credentials makes it essential for modern cybersecurity strategy.

Conclusion

In conclusion, authentication plays a critical role in ensuring the security of digital systems and networks. It is essential for organizations to implement robust authentication measures to protect sensitive information and prevent unauthorized access. Failure to do so can result in significant financial losses, reputational damage, and legal consequences for both the organization and its clients. Organizations must prioritize the implementation of multifactor authentication methods, such as biometric and token-based systems, to mitigate the risks associated with unauthorized access attempts. These methods provide an additional layer of security by requiring multiple forms of verification, making it much more difficult for hackers to infiltrate systems. Furthermore, organizations should regularly update and strengthen their authentication protocols to keep up with evolving cyber threats and ensure the ongoing protection of their digital assets. Regularly auditing and testing these protocols is crucial in identifying any vulnerabilities and addressing them promptly. By regularly assessing the effectiveness of authentication measures, organizations can proactively identify and rectify any weaknesses in their security systems, thus reducing the likelihood of successful cyberattacks. This proactive approach is essential for maintaining a strong cybersecurity posture and safeguarding sensitive information from potential breaches. It also demonstrates a commitment to the protection of customer and client data, enhancing trust and confidence in the organization's ability to secure information. This trust and confidence can ultimately lead to increased customer loyalty and a competitive advantage in the market. Organizations that prioritize authentication in cybersecurity can differentiate themselves from their competitors and attract customers who are increasingly concerned about data privacy and security. These customers are more likely to choose organizations that prioritize their security, leading to a larger customer base and potential business growth. This can result in increased revenue and success for the organization in the long term. As the threat landscape continues to evolve, organizations must remain vigilant in their commitment to authentication and cybersecurity to adapt to new challenges and protect their digital assets. By staying up to date with emerging technologies and best practices in authentication, organizations can stay one step ahead of cybercriminals and ensure the ongoing security of their systems and networks. This ongoing dedication to authentication will be crucial in maintaining the trust and loyalty of customers, as well as mitigating the potential risks and consequences associated with cyberattacks. In conclusion, authentication is not just a one-time implementation but an ongoing process that requires constant attention and adaptation. Organizations must continuously assess and improve their authentication methods to keep pace with the ever-changing cybersecurity landscape and protect their valuable digital assets. By doing so, they can minimize the potential for security breaches and maintain the integrity of their systems and networks. This commitment to authentication in cybersecurity is essential for the long-term success and sustainability of organizations in today's digital age. It is clear that the importance of authentication cannot be overstated, and organizations must make it a top priority in their cybersecurity strategies.

References

[1] Karim, Nader Abdel, et al. "Online Banking User Authentication Methods: A Systematic Literature Review." IEEE Access (2023). ieee.org

[2] Dastane, D. O. (2020). The effect of bad password habits on personal data breach. International Journal of Emerging Trends in Engineering Research, 8(10). academia.edu

[3] Thapa, C. & Camtepe, S. (2021). Precision health data: Requirements, challenges and existing techniques for data security and privacy. Computers in biology and medicine. arxiv.org

[4] Maltoni, D., Maio, D., Jain, A. K., & Feng, J. (2022). Fingerprint Sensing. Handbook of Fingerprint Recognition. HTML

[5] Rains, T. (2020). Cybersecurity Threats, Malware Trends, and Strategies: Learn to mitigate exploits, malware, phishing, and other social engineering attacks. 103.159.250.194

[6] Rathod, U., Sonkar, M., & Chandavarkar, B. R. (2020, July). An experimental evaluation on the dependency between one-way hash functions and salt. In 2020 11th International Conference on Computing, Communication and Networking Technologies (ICCCNT) (pp. 1–7). IEEE. researchgate.net

[7] Witman, P. D., & Mackelprang, S. (2022). The 2020 Twitter Hack—So Many Lessons to Be Learned. Journal of Cybersecurity Education, Research and Practice, 2021(2). ed.gov

PART II

Ransomware Basics and Prevention

Chapter 5: Introduction to Ransomware

- **What Is Ransomware?:** Defines ransomware and explains how it has evolved over time. Discusses the impact of ransomware on individuals and organizations.
- **Historical Overview:** Provides a timeline of significant ransomware attacks, highlighting key developments in ransomware tactics and defenses.

Chapter 6: Ransomware Lifecycle

- **Infection:** Outlines the common vectors ransomware uses to infect systems, such as phishing emails and exploited vulnerabilities.
- **Encryption:** Explains how ransomware encrypts files and systems, detailing the cryptographic principles behind ransomware attacks.
- **Extortion:** Discusses the methods attackers use to demand ransom from victims, including payment methods and the psychological tactics used to pressure payment.

PART II   RANSOMWARE BASICS AND PREVENTION

Chapter 7: Ransomware Prevention

- **Best Practices:** Offers a set of best practices for preventing ransomware infections, including regular software updates and user education.

- **Endpoint Security:** Delves into the technologies and strategies for securing endpoints against ransomware, including antivirus software and endpoint detection and response (EDR) systems.

- **Network Security:** Covers network-level defenses against ransomware, such as firewalls, intrusion detection systems (IDS), and secure network architectures.

CHAPTER 5

Introduction to Ransomware

Author:
Harshavardhan Nerella

What Is Ransomware?

Ransomware is defined as a type of malicious software attack where the attacker threatens to publish any data, or file, or compromise security or a computer system until the victim fulfills the ransom demands.

Ransomware is one of the most significant threats to individuals, businesses, and governments worldwide. It can lead to significant financial losses, disruption of operations, and potential exposure of sensitive information. [1]

Evolution of Ransomware

Ransomware can be traced back to 1989; believe it or not, it has been making its mark for more than 30 years.

Early forms of ransomware were inclusive of encrypting file names on infected systems and demanding ransom in exchange for restoration. These early forms were relatively rare and not particularly sophisticated.

CHAPTER 5   INTRODUCTION TO RANSOMWARE

They mostly used symmetric encryption, where the same key was used for the encryption and decryption of files. To counter this ransomware, attackers started using asymmetric encryption, which made it more difficult to recover files.

The ransomware was primarily distributed by email attachments or malicious links. Ransomware attacks evolved as the security measures improved and users became more cautious. The attackers adapted by employing more sophisticated delivery methods by exploiting software vulnerabilities, exploiting kits, and leveraging social engineering techniques such as phishing mails, malicious advertising, and fake software updates.

The emergence of more advanced ransomware was seen in the mid-2000s, when ransomware attacks leveraged a bit stronger encryption algorithms that were difficult to decrypt. The emergence of digital currencies like Bitcoin also made it nearly impossible for victims to recover their data without the decryption key, making it fairly easy for attackers to collect ransoms without being traced.

In the 2010s, the ransomware attacks grew exponentially, becoming more advanced, sophisticated, and targeted. Ransomware as a Service (RaaS) platform emerged where attackers, even with a small amount of technical expertise, launch ransomware attacks in exchange for a share of profits. This period saw the development of polymorphic ransomware, where malicious code was capable of change to avoid detection by antivirus software.

Recent trends of attackers have shifted toward attacks against businesses and organizations where higher ransoms can be demanded. These attacks usually involve intense exploration to identify valuable targets and exploit vulnerabilities. The strategy of "double extortion" is common where the attackers not only encrypt the file but also threaten the victim to release the sensitive stolen data if demands are not met.

CHAPTER 5   INTRODUCTION TO RANSOMWARE

Impact of Ransomware on Individuals and Organizations

Ransomware, a menacing form of cyberattack, casts a long shadow over both individuals and organizations, leaving behind a trail of devastation and disruption. For individuals, the impact of ransomware is deeply personal and often financially burdensome. Victims find themselves grappling with the anguish of losing access to cherished files, photos, and documents, held hostage by malicious actors demanding hefty ransoms. The emotional toll is profound, as the violation of privacy and sense of vulnerability linger long after the attack subsides. Furthermore, the specter of identity theft looms large, with sensitive personal information at risk of falling into the wrong hands, amplifying the distress experienced by those ensnared in the web of ransomware.

Organizations, on the other hand, face a multifaceted onslaught when confronted with ransomware attacks. Beyond the immediate financial losses incurred through ransom payments or remediation efforts, businesses grapple with operational disruptions that reverberate throughout their ecosystem. Critical systems are paralyzed, productivity plummets, and supply chains falter under the weight of encrypted data and compromised networks. The reputational damage inflicted by ransomware attacks can be equally crippling, eroding trust among customers, partners, and stakeholders. The fallout from such breaches extends beyond mere monetary losses, tarnishing the organization's brand and eroding its competitive standing in the marketplace.

Mitigation strategies serve as a bulwark against the rising tide of ransomware threats, offering individuals and organizations a lifeline amidst the chaos. For individuals, proactive measures such as regular data backups, robust cybersecurity practices, and heightened awareness can mitigate the risk of falling victim to ransomware. Similarly, organizations must fortify their defenses through employee training, security protocols,

CHAPTER 5   INTRODUCTION TO RANSOMWARE

and incident response planning to thwart ransomware attacks and minimize their impact. By investing in cybersecurity resilience, both individuals and organizations can bolster their ability to withstand the onslaught of ransomware and emerge stronger in the face of adversity.

Looking ahead, the landscape of ransomware continues to evolve, presenting new challenges and opportunities for individuals and organizations alike. Emerging trends, such as the proliferation of ransomware-as-a-service and the targeting of specific industries or sectors, underscore the need for vigilance and adaptability in the cybersecurity realm. Moreover, the convergence of artificial intelligence and machine learning holds promise in enhancing threat detection and response capabilities, offering a glimmer of hope in the ongoing battle against ransomware. By staying abreast of these developments and embracing innovative solutions, individuals and organizations can navigate the ever-changing threat landscape with confidence and resilience.

In conclusion, the impact of ransomware on individuals and organizations is profound and far-reaching, demanding a concerted effort to confront and mitigate its deleterious effects. From the personal anguish experienced by individual victims to the systemic disruptions faced by businesses, ransomware casts a long shadow over our digital world. However, through proactive measures, resilience, and innovation, we can turn the tide against ransomware, safeguarding our data, our livelihoods, and our collective future in an increasingly interconnected world.

Timeline of Prominent Ransomware Attacks

The first ever ransomware attack that was documented was after the World Health Organization AIDS conference in 1989 when 20,000 floppy disks were mailed to event attendees. The packaging suggested the disk has a questionnaire to determine the likelihood of someone contracting AIDS.

CHAPTER 5   INTRODUCTION TO RANSOMWARE

As no one had heard of ransomware before, the malware was able to make its way to victims' systems. It was dubbed the AIDS trojan, which used simple symmetric encryption to block users from accessing their files, and a message appeared on their screen demanding they mail $189 to a P.O. box in Panama in exchange for access to their files. This attack is popularly known as PC Cyborg. [2]

Despite its long history, ransomware attacks did not gain popularity till the 2000s when the internet was booming, and the internet had become a household community and a way of life.

The two most notable attacks at the start of the internet era were GPCode and Archievus.

The **GPCode** attack was in **2004** where it infected systems via malicious website links and phishing emails by using a custom encryption algorithm to encrypt the files on systems. As a ransom, the attacker demands $20 in exchange for a decryption key. Fortunately for victims, the encryption key was fairly straightforward to crack.

The **Archives** attack was in **2006** when the attackers understood the importance of strong encryption and had used the advanced 1024-bit RSA encryption code. It was a failure as attackers did not use different passwords to unlock the systems, which was discovered by victims, and Archievus fell out of favor. [2]

Locker Ransomware started in 2009 in Russia, and it was spread to the rest of the world. Locker ransomware mostly targeted mobile devices. In this attack, locker ransomware is generally installed when a victim visits a website with malicious code or malicious ads. It is run on the victim's device, and it creates a popup stating the device has been locked and the only way to unlock it is by paying the ransom, generally through gift cards or MoneyPak. The ransom note usually included suggestions on places where the gift card or MoneyPak vouchers can be purchased, making it easy for the victims to pay the ransom. On mobile devices, the ransomware is disguised as an app that usually does not raise any suspicion, e.g., calculator app. The app will in turn contain the malicious code, and when the app is opened, the phone gets locked.

Cryptolocker first emerged in September 2013 and wreaked havoc on countless individuals and organizations by encrypting files and demanding ransom payments in Bitcoin. Its reign of terror lasted until June 2014 when law enforcement agencies and cybersecurity experts disrupted its infrastructure, albeit leaving significant damage in its wake.

Cryptowall surfaced shortly after Cryptolocker in 2014, employing similar tactics to encrypt files and extort ransom payments. It continued to evolve and proliferate through exploit kits and malicious email campaigns, causing widespread financial losses and data breaches until its decline in 2015 following coordinated law enforcement actions and cybersecurity efforts.

Locky ransomware surged to prominence in early 2016, spreading rapidly through malicious spam emails containing infected attachments. Its peak activity occurred in May 2016, infecting hundreds of thousands of systems worldwide and causing extensive data loss and disruption before declining later that year due to improved detection and mitigation efforts by cybersecurity professionals.

Wannacry made headlines in May 2017 with its unprecedented global cyberattack, exploiting a vulnerability in Microsoft Windows systems. It infected over 200,000 computers across 150 countries within days, disrupting critical infrastructure, healthcare systems, and businesses worldwide. The attack prompted widespread alarm and led to increased efforts to patch vulnerabilities and enhance cybersecurity defenses.

NotPetya unleashed chaos in June 2017, masquerading as ransomware but functioning more as destructive malware. It spread rapidly through compromised software updates, affecting organizations globally, including multinational corporations and critical infrastructure providers. The attack caused billions of dollars in damages, highlighting the need for robust cybersecurity measures and international cooperation to combat cyber threats effectively.

CHAPTER 5 INTRODUCTION TO RANSOMWARE

DarkSide rose to infamy in 2021 as the perpetrator of high-profile ransomware attacks targeting critical infrastructure, corporations, and government agencies. Operating as a ransomware-as-a-service (RaaS) operation, DarkSide provided affiliates with sophisticated tools and techniques, resulting in a series of disruptive and lucrative attacks. Its activities culminated in the May 2021 Colonial Pipeline attack, drawing attention to the significant impact of ransomware on national security and economic stability.

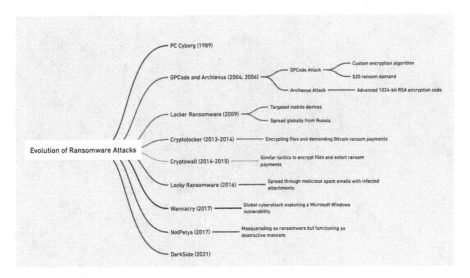

Image created by whimsical.com

References

[1] https://www.proofpoint.com/us/threat-reference/ransomware

[2] https://ransomware.org/what-is-ransomware/the-history-of-ransomware/

CHAPTER 6

Ransomware Lifecycle

Author:
Harshavardhan Nerella

Ransomware, a malicious software program, has become a widespread threat in the digital age. Ransomware can be imagined as a digital kidnapper encrypting valuable data and demanding a ransom for its release. Organizations and individuals must understand its lifecycle to develop defenses that are effective. This chapter discusses the lifecycle of ransomware attacks and delves deep into exploring the attacker's methods and the potential consequences at each stage.

Phase 1: Reconnaissance and Target Selection

Meticulous Planning: Scouting for Vulnerable Victims
The first phase lays the groundwork for the attack. Here, cybercriminals meticulously gather information about potential targets. They employ various techniques, including the following:

- **Social Engineering:** Attackers exploit human psychology through phishing emails, phone calls, or social media messages. These messages often masquerade as legitimate sources, tricking victims into clicking malicious links or downloading infected attachments.

- **Open-Source Intelligence (OSINT):** Cybercriminals leverage OSINT techniques to collect data available in the public domain. They analyze company websites, social media, professional networks, and other internet sources to gather information about an organization's employees, network structure, and potential security gaps.

- **Data Breaches:** Accessing and utilizing previously leaked or stolen information is another tactic. Attackers might buy credentials or exploit data from past breaches to infiltrate an organization's network, often bypassing initial security barriers without detection.

- **Advanced Persistent Threats (APTs):** In some cases, the reconnaissance activities are part of an advanced persistent threat (APTs), where attackers maintain a prolonged presence in the target's network to gather in-depth knowledge without being detected. This approach allows them to understand the network's intricacies, identify high-value assets, and plan their attack to cause maximum disruption.

Target Rich Environment: Why Your Organization Might Be at Risk
Several factors make organizations attractive targets:

- **Valuable Data:** Organizations dealing with critical financial records, intellectual property, or sensitive personal data are particularly appealing to attackers. Such data can be encrypted for ransom or even sold or leveraged for additional malicious activities.

- **Perceived Inability to Recover:** Attackers often target entities they perceive to be less capable of recovering from an attack without succumbing to ransom demands. Organizations lacking robust data backup and disaster recovery plans are seen as more likely to pay a ransom to restore their operations quickly.

- **Public Image:** High-profile organizations or those holding significant public trust may be targeted to maximize the impact of the attack. Attackers anticipate that the potential reputational damage and operational disruption will pressure these organizations into paying the ransom to resolve the crisis swiftly.

- **Geopolitical Motivations:** In some cases, ransomware attacks are motivated by geopolitical objectives. State-sponsored actors or cybercriminals with political agendas might target specific nations, industries, or entities to further their strategic interests or cause political disruption.

Phase 2: Initial Access

Breaching the Gates: Establishing a Foothold

Once a target is selected, attackers attempt to gain initial access to the victim's network. Common methods include:

- **Phishing Emails:** One of the most common infection vectors, phishing attacks involve sending emails that appear to be from legitimate sources but contain malicious links or attachments. When the recipient clicks on the link or downloads the attachment, malware is installed on their device.

CHAPTER 6 RANSOMWARE LIFECYCLE

Spear Phishing

A more targeted form of phishing, spear phishing involves emails that are customized to the recipient, making them more convincing. Attackers may use personal information gathered during the reconnaissance phase to craft these deceptive messages.

- **Exploiting Vulnerabilities:** Attackers often exploit vulnerabilities in software to inject ransomware. These vulnerabilities can be in operating systems, applications, or even in hardware firmware.

Unpatched Software

One common target is software that has not been updated or patched. Attackers take advantage of known vulnerabilities that have yet to be fixed on the target's system.

- **Watering Hole Attacks:** Attackers compromise legitimate websites frequently visited by the target organization. When employees visit these sites, malware is downloaded onto their devices.

- **Malvertising:** Malvertising involves injecting malicious code into legitimate advertising networks. When a user clicks on the ad or even just visits a website displaying the ad, the code can execute and install ransomware.

  Drive-by downloads occur when a user unknowingly downloads ransomware simply by visiting a compromised website. These downloads can happen without any interaction from the user, exploiting vulnerabilities in web browsers or plugins.

- **Remote Desktop Protocol (RDP) Exploits:** RDP is a popular tool for remote administration, but it's also a common vector for ransomware attacks. Attackers exploit weak or stolen credentials to gain remote access to systems and deploy ransomware.

- **Supply Chain Attacks:** In a supply chain attack, attackers target less-secure elements in the supply chain to reach the ultimate target. By compromising a trusted third-party vendor or software provider, attackers can deliver ransomware to multiple victims at once.

The Human Factor: Why Employees Remain Vulnerable

Social engineering tactics often succeed because they prey on human trust and inattention. Organizations can mitigate this risk by educating employees on cybersecurity best practices, such as:

- **Phishing Awareness Training:** Equipping employees to identify and avoid phishing attempts.

- **Strong Password Policies:** Enforcing the use of complex passwords and regular password changes.

- **Caution Toward Suspicious Links and Attachments:** Educating employees to be wary of unsolicited emails and attachments, even if they appear to come from a trusted source.

Phase 3: Lateral Movement and Privilege Escalation

Expanding the Reach: Moving Within the Network

After gaining initial access, attackers aim to move laterally across the network, compromising additional devices and accounts. Techniques employed at this stage include:

- **Exploiting Active Directory:** In many corporate environments, Active Directory (AD) is the cornerstone of network management, providing authentication and authorization services. Attackers often target AD to gain elevated privileges and access critical resources. By exploiting AD vulnerabilities, attackers can escalate their privileges, manipulate policies, and move freely across the network.

- **Pass-the-Hash Attacks:** These attacks are a prevalent method for lateral movement, where attackers capture password hashes and use them to authenticate to other systems within the network. This technique allows attackers to bypass password protection mechanisms and gain access to multiple accounts and systems without needing the plaintext passwords.

- **Remote Access Tools:** Legitimate remote access tools, which are essential for network administration and support, can be co-opted by attackers to facilitate lateral movement. By exploiting these tools, attackers can remotely control other systems on the network, expand their foothold, and deploy ransomware on additional targets.

- **Escalation of Privileges:** Ransomware often tries to gain higher system privileges to access more resources and data. This can involve exploiting system vulnerabilities, stealing credentials, or using hardcoded credentials often found in poorly secured systems.

- **Evading Detection:** Modern ransomware is designed to evade detection by antivirus and other security software. Techniques include obfuscating the code, changing its behavior, or even disabling security measures before executing the malicious payload.

Why Lateral Movement Matters: Expanding the Attack Surface
Lateral movement allows attackers to:

- Locate and encrypt valuable data across the network. By moving laterally across the network, attackers can identify and encrypt critical data stored on various systems, thereby magnifying the potential disruption and increasing the pressure on the organization to comply with the ransom demands.

- Disable security controls and backups. Attackers often target security tools and backup systems during lateral movement. By disabling these defenses, they aim to prevent the detection of the ransomware and hinder the organization's ability to recover from the attack without paying the ransom.

- Establish persistence within the network, ensuring continued access even if the initial infection point is discovered. Ensuring that their access remains uninterrupted, even if the initial entry points are detected and remediated, is a key goal during this phase. Establishing persistence mechanisms across the network allows attackers to retain access and control, posing a prolonged threat to the organization.

Phase 4: Deployment of Ransomware Payload

The Moment of Truth: Unleashing Encryption

Once attackers have established a foothold within the network and escalated privileges, they deploy the ransomware payload. The ransomware program then begins to:

- **Scan for Target Files:** The ransomware employs sophisticated scanning algorithms to identify files based on extensions, locations, or other attributes that suggest their importance. It aims to locate data that, when encrypted, would cause substantial operational disruption or financial loss, thereby increasing the pressure on the victim to pay the ransom.

- **Encrypt Target Files:** Using advanced encryption algorithms, the ransomware encrypts the identified files. This encryption is typically strong, using algorithms such as AES or RSA, which are practically impossible to break without the unique decryption key. The encryption process is designed to be rapid and efficient, minimizing the window for intervention before substantial data is rendered inaccessible.

- **Leave a Ransom Note:** The attackers typically leave a ransom.

CHAPTER 6   RANSOMWARE LIFECYCLE

The ransomware note serves a critical purpose for attackers. It typically includes:

- **The Extent of the Attack:** The note details the scope of the attack, often specifying the number of files or systems affected. This information is meant to underscore the severity of the situation, prompting a sense of urgency in the victim.

- **The Ransom Demand:** The demanded ransom amount is usually articulated in the note, often in a cryptocurrency like Bitcoin to maintain anonymity. The sum demanded can vary widely, influenced by the attackers' assessment of the victim's financial wherewithal and the critical nature of the encrypted data.

- **Payment Instructions:** Clear instructions are provided on how to execute the payment, including the cryptocurrency wallet address and possibly a deadline to increase the pressure. This part of the note is crucial for the attackers as it dictates the terms of their financial gain from the operation.

- **Threat of Data Leak:** To amplify the pressure, the note may threaten the public release or sale of the encrypted data if the ransom is not paid within a certain timeframe. This threat targets not just the operational capacity of the victim but also their reputation and the privacy of any stakeholders involved.

CHAPTER 6   RANSOMWARE LIFECYCLE

The Peril of Paying the Ransom: A Risky Gamble
While some organizations may be tempted to pay the ransom to regain access to their data quickly, it's a risky decision. Here's why:

- **No Guarantee of Decryption:** There is a significant risk that even after payment, the attackers may not provide the decryption key or that the key provided may not work as intended, leaving the victim with a financial loss and ongoing data inaccessibility.

- **Encourages Future Attacks:** Succumbing to the ransom demands emboldens the attackers and the broader ransomware community by validating the effectiveness of their tactics, potentially leading to further attacks against the same victim or others.

- **Legal and Regulatory Implications:** Engaging with the attackers by paying the ransom can have legal ramifications, particularly in regions or industries where such payments are regulated or prohibited. Moreover, it may involve complex ethical considerations, especially if the payment indirectly funds illicit activities or further criminal endeavors.

Phase 5: Encryption and Impact

The Digital Lockdown: Feeling the Bite
This phase marks the culmination of the attacker's efforts. The ransomware program goes into overdrive, encrypting target files across the network. The impact on the victim organization can be devastating:

CHAPTER 6   RANSOMWARE LIFECYCLE

- **Data Inaccessibility:** The encryption of files renders them inaccessible, effectively locking out users and halting any business operation reliant on that data. For organizations that depend heavily on digital data for their day-to-day operations, this can bring business to a standstill. Departments like finance, human resources, customer service, and IT can find themselves in a chokehold, unable to access the data they need to function.

- **Productivity Loss:** The ripple effect of data inaccessibility is a significant drop in productivity. Employees are unable to perform their duties effectively, critical processes are delayed, and the overall workflow of the organization is disrupted. This loss extends beyond the employees directly affected by the encrypted files, impacting various facets of the organization.

- **Financial Losses:** The financial ramifications of a ransomware attack are multifaceted. There's the immediate financial strain from the ransom demand itself, which can be substantial. Beyond that, the organization suffers from operational downtime, leading to lost revenue, potential penalties for breached contracts or service level agreements, and the costs associated with remediation efforts to restore the impacted systems.

- **Reputational Damage:** An organization's reputation takes a significant hit when it falls victim to a ransomware attack. Customers, partners, and stakeholders may lose trust in the organization's ability

CHAPTER 6 RANSOMWARE LIFECYCLE

to safeguard data, impacting customer loyalty and potentially leading to a loss of business. If the attack becomes public knowledge, the organization might face scrutiny from the media, industry regulators, and the general public, further tarnishing its reputation.

Phase 6: Extortion and Communication

Negotiation Tactics: A Digital Standoff

After the encryption is complete, attackers establish communication channels with the victim. This phase involves:

- **Negotiation Attempts:** Faced with the prospect of significant losses, victims may engage with the attackers, seeking to negotiate the ransom amount. In some cases, organizations attempt to buy time, reduce the payment, or verify the possibility of data recovery. However, engaging with the attackers is fraught with risks and uncertainties, as there is no guarantee of ethical conduct from the attackers' side.

- **Pressure Tactics:** Attackers often employ a range of pressure tactics to compel the organization to comply with their demands swiftly. This might include setting deadlines for payment, after which the ransom will increase or the encrypted data will be permanently locked or deleted. Additionally, if the attackers have exfiltrated data before encryption, they may threaten to release sensitive information publicly or sell it on dark web markets, leveraging the threat of reputational damage or legal repercussions to force the organization's hand.

- **Seeking External Assistance:** In this high-pressure situation, many organizations turn to external experts for help. Cybersecurity firms specialize in ransomware mitigation and may assist in assessing the situation, negotiating with the attackers, or restoring systems. Law enforcement agencies can provide guidance and support, although their involvement might not always lead to the recovery of the encrypted data.

Communication Phase: Navigating a Minefield

This phase of the attack is laden with critical decisions and intense negotiations. Organizations must navigate this challenging terrain with a strategic approach, balancing the urgency to restore operations with the risks of capitulating to the attackers' demands.

- **Assessing Communication:** Any communication with the attackers provides insights into their demands, deadlines, and potential willingness to reduce the ransom or provide proof of the decryptability of the data. However, every interaction should be handled with caution, understanding the potential legal implications and the risk of exacerbating the situation.

- **Strategic Response:** The organization must strategize its response, weighing the pros and cons of complying with the ransom demand against the potential for data recovery through other means. This includes considering the impact of the decision on future security, the precedent it sets, and the message it sends to potential future attackers.

- **Collaboration and Transparency:** Engaging with stakeholders, legal counsel, cybersecurity experts, and law enforcement can provide a multi-faceted perspective on the situation, aiding in the formulation of a response that aligns with legal requirements, ethical considerations, and the organization's best interests.

CHAPTER 7

Ransomware Prevention

Author:

Harshavardhan Nerella

In the battle against ransomware, implementing effective prevention measures is paramount to safeguarding your organization's data and infrastructure. This chapter outlines crucial best practices to mitigate the risk of ransomware attacks.

Regular Software Updates and Patch Management

In the ever-evolving landscape of cybersecurity threats, the exploitation of software vulnerabilities stands as one of the most favored tactics employed by ransomware actors. These adversaries capitalize on unpatched systems and outdated software to infiltrate networks, encrypt data, and extort hefty ransoms from their victims. Therefore, maintaining a rigorous regimen of software updates and robust patch management is imperative for fortifying the digital defenses of organizations.

CHAPTER 7   RANSOMWARE PREVENTION

Why is this important?

- **Exploitation of Known Vulnerabilities:** Cybercriminals exploit known vulnerabilities in software and operating systems to orchestrate ransomware attacks. These vulnerabilities serve as gateways for unauthorized access, enabling attackers to penetrate networks and execute malicious payloads.

- **Preying on Unpatched Systems:** Attackers prey on organizations with lax patch management practices, as these present lucrative opportunities for exploitation. Unpatched vulnerabilities provide adversaries with the foothold they need to launch ransomware campaigns, resulting in data encryption and extortion demands.

Best Practices:

- **Establish Patch Management Procedures:** Develop comprehensive patch management procedures that delineate the systematic identification, testing, and deployment of patches across all software and systems within the organization. These procedures should encompass clear delineations of roles, responsibilities, and escalation paths to ensure seamless patch deployment.

- **Automate Patch Deployment:** Embrace the power of automation by leveraging advanced patch management tools to streamline the deployment of updates and patches. Automation not only expedites the patching process but also minimizes the risk of human error inherent in manual patch deployment methods.

- **Prioritize Critical Vulnerabilities:** Adopt a risk-based approach to vulnerability management by prioritizing the remediation of critical vulnerabilities that pose the greatest threat to organizational security. Allocate resources judiciously to address vulnerabilities with the highest likelihood of exploitation by ransomware actors.

- **Continuous Vulnerability Monitoring:** Implement robust vulnerability monitoring mechanisms to proactively identify emerging threats and vulnerabilities affecting organizational assets. Continuous monitoring empowers organizations to stay one step ahead of attackers by swiftly identifying and mitigating vulnerabilities before they can be leveraged in ransomware attacks.

- **Maintain Inventory of Software Assets:** Cultivate an accurate inventory of all software applications and systems deployed across the organization's infrastructure. A comprehensive inventory serves as the cornerstone of effective patch management, ensuring that no critical systems or software are overlooked during the patching process.

- **Test Patches Before Deployment:** Prior to deploying patches in a production environment, conduct thorough testing to assess the compatibility and stability of the updates. Rigorous testing minimizes the risk of unintended consequences or disruptions resulting from patch deployment, thereby safeguarding the integrity of organizational systems.

CHAPTER 7   RANSOMWARE PREVENTION

Employee Education and Training

While technological solutions are integral to ransomware prevention, the human element remains a critical factor in defending against these sophisticated attacks. Employees are often the first line of defense against ransomware threats, making comprehensive education and training initiatives essential components of any cybersecurity strategy.

Why is this important?

- **Human Error as a Vulnerability:** Ransomware attackers frequently exploit human vulnerabilities, such as unsuspecting employees clicking on malicious links or downloading infected attachments. Educating employees about the dangers of ransomware and imparting essential cybersecurity skills can help mitigate these risks.

- **Enhancing Cyber Awareness:** By fostering a culture of cyber awareness within the organization, employees become more vigilant and better equipped to recognize potential ransomware threats. Educated employees serve as an additional layer of defense, helping to detect and thwart attacks before they can inflict significant damage.

Best Practices:

- **Cybersecurity Awareness Training:** Implement regular cybersecurity awareness training sessions to educate employees about the various forms of ransomware threats, common attack vectors, and best practices for mitigating risks. These sessions should cover topics such as phishing awareness, safe browsing habits, and email hygiene.

CHAPTER 7   RANSOMWARE PREVENTION

- **Simulated Phishing Exercises:** Conduct simulated phishing exercises to provide employees with hands-on experience in identifying and responding to phishing attempts. These exercises simulate real-world scenarios, allowing employees to practice discerning legitimate emails from phishing attempts and reinforcing cybersecurity awareness.

- **Role-Based Training:** Tailor training programs to the specific roles and responsibilities of employees within the organization. Different departments may face unique cybersecurity challenges, and targeted training ensures that employees receive relevant and actionable guidance to safeguard against ransomware threats.

- **Reporting Procedures:** Establish clear reporting procedures for employees to follow in the event of a suspected ransomware attack or security incident. Encourage a culture of transparency and accountability, empowering employees to report suspicious activities promptly to the appropriate IT or security personnel for investigation.

- **Regular Updates and Refreshers:** Cyberthreats evolve rapidly, necessitating regular updates and refreshers to keep employees informed about the latest ransomware trends and mitigation strategies. Incorporate ongoing education initiatives to reinforce key concepts and ensure that employees remain vigilant against emerging threats.

- **Promote a Security-Conscious Culture:** Foster a security-conscious culture within the organization by recognizing and rewarding employees who demonstrate exemplary cybersecurity practices. Encourage open communication about cybersecurity concerns and promote collaboration between departments to strengthen the collective defense against ransomware attacks.

Implementing Robust Endpoint Security Solutions

Endpoint security is critical in defending against ransomware attacks, as endpoints serve as the primary targets for infiltration and data encryption. Implementing robust endpoint security solutions is essential for detecting and thwarting ransomware threats before they can wreak havoc on organizational networks.

Why is this important?

- **Endpoint Vulnerabilities:** Endpoints, including desktops, laptops, mobile devices, and servers, are often the entry points for ransomware attacks. These devices are vulnerable to exploitation through various attack vectors, such as malicious emails, infected attachments, and compromised websites.

- **Detection and Response:** Effective endpoint security solutions provide advanced detection and response capabilities to identify ransomware activity in real time. By detecting ransomware behavior early, organizations can take prompt action to contain the threat and prevent further damage.

Best Practices:

- **Antivirus Software**: Deploy reputable antivirus software across all endpoints to detect and block known malware, including ransomware variants. Ensure that antivirus definitions are regularly updated to protect against the latest threats.

- **Endpoint Detection and Response (EDR) Systems**: Implement EDR systems to monitor endpoint activity for signs of ransomware behavior, such as file encryption and unauthorized access attempts. EDR solutions offer advanced threat detection capabilities and facilitate rapid incident response.

- **Behavioral Analysis**: Leverage behavioral analysis techniques to identify anomalous endpoint behavior indicative of ransomware activity. Behavioral analysis solutions analyze endpoint activity patterns to detect deviations from normal behavior and trigger alerts for further investigation.

- **Application Whitelisting**: Utilize application whitelisting to restrict the execution of unauthorized software on endpoints. By only allowing approved applications to run, organizations can prevent the execution of ransomware and other malicious programs.

- **Endpoint Firewall**: Implement endpoint firewalls to monitor and control network traffic to and from endpoints. Endpoint firewalls help block unauthorized network communications associated with ransomware attacks, such as command-and-control traffic.

- **Device Encryption**: Enable device encryption on endpoints to protect sensitive data from unauthorized access in the event of device theft or loss. Encryption safeguards data stored on endpoints, rendering it inaccessible to ransomware attackers even if the device is compromised.

Data Backup and Disaster Recovery Planning

Data backup and disaster recovery planning are indispensable components of ransomware prevention strategies, offering organizations a lifeline in the event of a successful ransomware attack. By implementing robust backup solutions and comprehensive disaster recovery plans, organizations can mitigate the impact of ransomware incidents and ensure the timely restoration of critical data and systems.

Why is this important?

- **Ransomware Resilience**: Data backup and disaster recovery planning serve as the last lines of defense against ransomware attacks, providing organizations with the ability to restore encrypted or inaccessible data without capitulating to ransom demands. In the absence of effective backup measures, organizations risk permanent data loss and operational disruptions, which can have profound financial and reputational repercussions.

- **Business Continuity**: Effective backup and disaster recovery measures are essential for maintaining business continuity in the face of ransomware incidents. By swiftly recovering from ransomware

CHAPTER 7   RANSOMWARE PREVENTION

attacks, organizations can minimize downtime, mitigate financial losses, and preserve customer trust and satisfaction. Rapid data restoration enables organizations to resume critical business operations and minimize the impact of ransomware-induced disruptions on productivity and revenue generation.

Best Practices:

- **Regular Data Backups**: Implement a regular schedule for backing up critical data and systems to secure, off-site locations. Automated backup solutions can streamline the backup process and ensure that data is consistently protected against ransomware threats. Organizations should prioritize the backup of essential data assets, including customer records, financial information, intellectual property, and operational data.

- **Multiple Backup Copies**: Maintain multiple backup copies of important data to guard against data loss due to backup failures or corruption. Storing backup copies in diverse locations, such as off-site data centers or cloud repositories, enhances redundancy and resilience, minimizing the risk of simultaneous data loss in the event of a ransomware attack or natural disaster. Redundant backup copies serve as insurance against unforeseen contingencies, enabling organizations to recover swiftly from ransomware incidents with minimal data loss.

- **Immutable Backup Storage**: Utilize immutable backup storage solutions that prevent unauthorized modification or deletion of backup data. Immutable storage technologies, such as write-once, read-many

(WORM) storage, protect backup data from tampering or alteration by ransomware attackers. By enforcing strict access controls and cryptographic protections, immutable backup storage ensures the integrity and availability of backup data, even in the face of sophisticated ransomware attacks.

- **Regular Backup Testing**: Conduct regular testing and validation of backup systems and processes to verify the integrity and recoverability of backup data. Testing ensures that backup solutions are capable of effectively restoring data in the event of a ransomware incident or other disasters. Organizations should simulate various ransomware scenarios, including data corruption, encryption, and deletion, to assess the efficacy of backup recovery procedures and identify potential gaps or vulnerabilities in backup infrastructure.

- **Disaster Recovery Planning**: Develop comprehensive disaster recovery plans that outline procedures for responding to ransomware incidents and other catastrophic events. Disaster recovery plans should encompass strategies for data restoration, system recovery, and business continuity measures. Organizations should define clear roles, responsibilities, and communication protocols to facilitate a coordinated and effective response to ransomware attacks. Regular drills and tabletop exercises can help validate disaster recovery plans and ensure that personnel are prepared to execute response procedures effectively in high-pressure situations.

CHAPTER 7 RANSOMWARE PREVENTION

- **Incident Response Protocols**: Establish incident response protocols to guide the organization's response to ransomware attacks. Clearly define roles, responsibilities, and escalation procedures to ensure a coordinated and effective response to ransomware incidents. Organizations should establish communication channels for reporting and triaging ransomware incidents, enabling rapid detection, containment, and mitigation of threats. Incident response teams should be equipped with the necessary tools, resources, and authority to investigate ransomware incidents thoroughly and implement remediation measures to minimize the impact on organizational operations and data assets.

Network Segmentation and Access Controls

Network segmentation and access controls are pivotal components of ransomware prevention strategies, serving as barriers to limit the lateral movement of ransomware within organizational networks. By partitioning networks into distinct zones and enforcing stringent access controls, organizations can minimize the propagation of ransomware and contain its impact on critical systems and data.

Why is this important?

- **Limiting Ransomware Spread**: Ransomware attacks often rely on lateral movement to propagate within organizational networks, leveraging compromised endpoints to infect additional systems and escalate privileges. Network segmentation and access controls restrict the movement of ransomware by segmenting networks into isolated zones and enforcing granular access policies.

- **Containment and Mitigation**: In the event of a ransomware incident, network segmentation and access controls facilitate containment and mitigation efforts by confining the impact of ransomware to specific network segments. By isolating infected systems and preventing unauthorized access to critical resources, organizations can minimize the scope of ransomware-induced disruptions and expedite recovery efforts.

Best Practices:

- **Segmentation of Network Infrastructure**: Divide network infrastructure into discrete segments or zones based on logical or functional criteria, such as departmental boundaries, data sensitivity levels, or operational requirements. Segmenting networks isolates critical systems and data assets from less secure areas, limiting the potential blast radius of ransomware attacks and containing their impact.

- **Implementing Firewalls and Access Controls**: Deploy firewalls and access control mechanisms to enforce traffic filtering and restrict communication between network segments. Configure firewall rules to allow only authorized communication flows and block suspicious or unauthorized network traffic associated with ransomware activities. Implement network access controls, such as VLANs, subnetting, and role-based access control (RBAC), to enforce least privilege principles and limit access to sensitive resources.

- **Network Monitoring and Intrusion Detection:** Implement network monitoring solutions and intrusion detection systems (IDS) to detect anomalous network activity indicative of ransomware infections or unauthorized access attempts. Continuous monitoring enables organizations to identify and respond to ransomware threats in real time, allowing for swift containment and remediation actions to mitigate the impact of attacks.

- **Endpoint Isolation and Quarantine:** Implement endpoint isolation and quarantine measures to contain ransomware-infected devices and prevent further spread within the network. Isolate compromised endpoints from the rest of the network to prevent ransomware from propagating to additional systems and encrypting data. Leverage network access control solutions to automatically quarantine infected devices and restrict their network connectivity until they can be remediated.

- **Segmentation of Privileged Access:** Segment and restrict access to privileged accounts and administrative interfaces to minimize the risk of ransomware attacks targeting critical infrastructure and administrative systems. Implement strict access controls and authentication mechanisms to prevent unauthorized access to privileged accounts and limit the impact of ransomware incidents on essential network components.

- **Regular Security Audits and Assessments**: Conduct regular security audits and assessments to evaluate the effectiveness of network segmentation and access controls in mitigating ransomware risks. Identify and remediate vulnerabilities or misconfigurations that could undermine the integrity of network segmentation or compromise access control mechanisms. Continuously monitor and adjust segmentation policies to adapt to evolving threats and organizational requirements.

Email and Web Security Measures

Email and web security measures are pivotal in ransomware prevention, as cybercriminals often exploit these channels to deliver malicious payloads and infiltrate organizational networks. By implementing robust email and web security controls, organizations can thwart ransomware attacks at the point of entry, preventing the compromise of endpoints and the encryption of critical data.

Why is this important?

- **Primary Attack Vectors**: Email and web-based attacks are among the primary vectors used by ransomware actors to distribute malicious payloads and lure unsuspecting users into clicking on malicious links or downloading infected attachments. Strengthening email and web security defenses is crucial for blocking ransomware threats at their inception and preventing their propagation within organizational networks.

- **Human Element**: Ransomware attacks often exploit human vulnerabilities, such as lack of awareness or vigilance, to gain a foothold within organizations. Email and web security measures help mitigate these risks by detecting and blocking malicious content before it reaches end users, reducing the likelihood of successful ransomware infections.

Best Practices:

- **Email Filtering and Anti-phishing Measures**: Deploy email filtering solutions equipped with advanced anti-phishing capabilities to detect and block phishing emails containing ransomware payloads or malicious links. Implement sender authentication mechanisms, such as Sender Policy Framework (SPF), DomainKeys Identified Mail (DKIM), and Domain-based Message Authentication, Reporting, and Conformance (DMARC), to verify the authenticity of email senders and mitigate email spoofing attacks.

- **Content and Attachment Scanning**: Scan email attachments and embedded content for signs of malicious activity, such as executable files or suspicious macros. Use heuristic analysis and machine learning algorithms to identify potentially malicious content and quarantine suspicious emails before they reach end users' inboxes. Educate users about the dangers of opening unsolicited email attachments or clicking on links from unknown or untrusted sources.

- **URL Filtering and Web Security Gateways**: Implement URL filtering and web security gateways to block access to malicious websites hosting ransomware payloads or exploit kits. Use threat intelligence feeds and reputation-based filtering to categorize and block known malicious URLs, preventing users from inadvertently visiting compromised websites or downloading malicious content. Configure web security gateways to enforce browsing policies and restrict access to high-risk or non-business-related websites.

- **Secure Email Encryption and Data Loss Prevention**: Encrypt sensitive email communications and implement data loss prevention (DLP) controls to prevent the unauthorized transmission of sensitive data via email. Use email encryption technologies, such as Transport Layer Security (TLS) and Pretty Good Privacy (PGP), to secure email communications in transit and protect sensitive information from interception or unauthorized access. Implement DLP policies to monitor outbound email traffic and prevent the inadvertent disclosure of sensitive data in email messages.

- **User Awareness Training**: Provide comprehensive cybersecurity awareness training to educate users about the risks of ransomware attacks via email and the web. Train users to recognize common ransomware lures and phishing tactics, such as urgent requests for personal information or unsolicited email attachments from unknown senders. Empower users to report suspicious emails or web links to the IT or security team for further investigation and remediation.

- **Email Authentication and Spoofing Prevention**: Implement email authentication protocols, such as Domain-based Message Authentication, Reporting, and Conformance (DMARC), to prevent email spoofing and domain impersonation attacks. Configure DMARC policies to specify how email servers should handle messages that fail authentication checks, such as quarantine or reject. Monitor DMARC reports and adjust policies as needed to improve email authentication and reduce the risk of domain spoofing attacks.

Developing a Comprehensive Prevention Strategy

A comprehensive ransomware prevention strategy integrates multiple layers of defense mechanisms and proactive measures to mitigate the risk of ransomware attacks and minimize their impact on organizational operations and data assets. By adopting a holistic approach to ransomware prevention, organizations can strengthen their resilience and enhance their ability to detect, respond to, and recover from ransomware incidents effectively.

Why is this important?

- **Multi-faceted Defense**: Ransomware attacks are constantly evolving, requiring organizations to deploy a diverse array of prevention measures to address emerging threats and vulnerabilities effectively. A comprehensive prevention strategy combines technological controls, user education, incident response protocols, and risk management practices to create a robust defense posture against ransomware attacks.

CHAPTER 7   RANSOMWARE PREVENTION

- **Risk Mitigation:** Developing a comprehensive prevention strategy enables organizations to identify and mitigate the risk factors associated with ransomware attacks, such as outdated software, insecure configurations, and human error. By addressing these risk factors proactively, organizations can reduce their susceptibility to ransomware threats and minimize the likelihood of successful attacks.

Best Practices:

- **Risk Assessment and Vulnerability Management:** Conduct regular risk assessments and vulnerability scans to identify weaknesses and security gaps that could be exploited by ransomware attackers. Prioritize remediation efforts based on the severity and likelihood of exploitation of identified vulnerabilities, focusing on critical systems and high-risk assets.

- **Defense-in-Depth Architecture:** Implement a defense-in-depth architecture that layers multiple security controls and safeguards to protect against ransomware attacks at different stages of the cyber kill chain. Combine endpoint security solutions, network segmentation, email filtering, user awareness training, and incident response capabilities to create a multilayered defense posture that mitigates the risk of ransomware infiltration and propagation.

- **Incident Response Planning and Preparedness:** Develop comprehensive incident response plans that outline procedures for detecting, analyzing, and responding to ransomware incidents effectively.

Define roles, responsibilities, and communication channels to facilitate a coordinated and timely response to ransomware attacks. Conduct regular tabletop exercises and incident response drills to test the effectiveness of response procedures and improve incident handling capabilities.

- **Continuous Monitoring and Threat Intelligence:** Implement continuous monitoring and threat intelligence capabilities to detect and respond to ransomware threats in real time. Leverage security information and event management (SIEM) systems, intrusion detection systems (IDS), and threat intelligence feeds to identify anomalous behavior, indicators of compromise (IOCs), and emerging ransomware threats. Stay abreast of the latest ransomware trends, tactics, and techniques to adapt prevention strategies accordingly and anticipate future threats.

- **User Education and Awareness Training:** Provide comprehensive cybersecurity education and awareness training to empower users to recognize and mitigate ransomware threats effectively. Train employees on safe computing practices, phishing awareness, social engineering tactics, and incident reporting procedures. Foster a culture of security awareness and accountability to ensure that all members of the organization play an active role in ransomware prevention efforts.

- **Regular Testing and Validation:** Conduct regular testing, validation, and review of ransomware prevention controls and procedures to assess their effectiveness and identify areas for improvement. Perform penetration testing, red team exercises, and security audits to evaluate the resilience of defense mechanisms and validate the organization's readiness to withstand ransomware attacks. Use testing results and lessons learned to refine prevention strategies and enhance the organization's overall security posture.

PART III

Ransomware Detection and Response

Chapter 8: Early Detection Techniques

- **Behavior Analysis:** Introduces the concept of behavior analysis in detecting ransomware, focusing on how anomalous behavior can indicate an infection.
- **Anomaly Detection:** Explains methods for detecting anomalies in system and network behavior that may signal a ransomware attack.

Chapter 9: Incident Response

- **Developing an Incident Response Plan:** Guides readers through the creation of an incident response plan tailored to ransomware threats.
- **Execution of Incident Response:** Discusses the steps involved in responding to a ransomware attack, from identification and containment to eradication and recovery.

PART III   RANSOMWARE DETECTION AND RESPONSE

Chapter 10: Threat Intelligence

- **Leveraging Threat Feeds:** Explores how threat intelligence feeds can be used to stay informed about emerging ransomware threats and vulnerabilities.

- **Collaborative Threat Intelligence Sharing:** Discusses the importance of sharing threat intelligence within the cybersecurity community and how collaborative efforts can enhance defense against ransomware.

CHAPTER 8

Early Detection Techniques

Author:
Dhruv Seth

In the realm of cybersecurity, if outright ransomware prevention is unattainable, the subsequent priority becomes its swift detection. Despite being equipped with the latest antivirus solutions and comprehensive defense mechanisms, a staggering 85% of organizations afflicted by ransomware were found to have these protective measures in place, yet still fell prey to attacks. This chapter delves into optimal strategies for the early detection of ransomware, aiming to halt its progression and minimize the resultant harm.

Unveiling the Challenge of Ransomware Detection

The paradox that ransomware often breaches organizations fortified with traditional cybersecurity defenses is bewildering to many. Such entities typically maintain updated antivirus tools, firewalls, and other critical safeguards, which theoretically should provide robust protection against digital threats. The chapter explores the multifaceted reasons behind the successful infiltration of seemingly secure networks by ransomware operators.

CHAPTER 8   EARLY DETECTION TECHNIQUES

Key hurdles include the inherent limitations of antivirus and endpoint detection solutions, which, despite marketing claims, cannot guarantee absolute accuracy. The rapid evolution of malware, coupled with the agility of cybercriminals in altering malware signatures, significantly strains the capacity of signature-based detection systems. Moreover, the imperfect implementation of security measures, the strategic disabling of protective tools by ransomware, and the sophisticated use of legitimate system tools for malicious purposes further complicate detection efforts.

Strategies for Enhanced Detection

Addressing the detection of ransomware necessitates a blend of traditional and advanced methodologies. Beyond the standard cybersecurity arsenal, this section advocates for a proactive stance on detection, emphasizing the importance of recognizing and acting upon the nuanced indicators of ransomware activity.

- **Security Awareness Training:** Training people about online safety is like giving them a shield to protect themselves and the company from cyberattacks, especially ransomware. By learning these smarts together, we can make the company much less likely to get attacked and keep our information and reputation safe.

CHAPTER 8 EARLY DETECTION TECHNIQUES

Elevating Cybersecurity Literacy

- **Cybersecurity Literacy:** This refers to the knowledge and understanding that administrators and end-users have about cybersecurity threats and safe practices online. It encompasses a broad range of topics, from recognizing phishing emails to understanding the importance of regular software updates.

- **Both Administrators and End-Users:** Training isn't limited to the IT department. While administrators might need more technical training, end-users must understand basic cybersecurity practices because they often serve as the entry point for attackers.
- **Crucial for Organizational Security:** Elevating literacy is not just beneficial; it's essential. With the increasing sophistication of cyberthreats, everyone in an organization needs to be aware of how to protect themselves and the organization.

Familiarizing with the Hallmarks of Ransomware

- **Ransomware:** A type of malicious software designed to block access to a computer system or files until a sum of money is paid. It's a prevalent and destructive type of cyberattack.
- **Hallmarks of Ransomware:** These are the indicators or warning signs of ransomware, such as suspicious email attachments, unexpected software behavior, or unusual file extensions. Training helps individuals recognize these signs early.

Empowering the Workforce

- **Identify and Report Suspicious Activities:** Training equips employees with the knowledge to not only recognize potential threats but also understand the channels through which they should report these activities.

- **Vital First Line of Defense:** Employees are on the front lines. By recognizing and reporting suspicious activities, they can prevent many attacks from succeeding. This proactive approach is significantly more cost effective than addressing the consequences of a successful cyberattack.

Implementation and Benefits

Interactive Training Sessions: Effective training should engage participants, encourage questions, and include practical exercises that mimic real-life scenarios.

Regular Updates and Refreshers: Cyberthreats evolve rapidly; thus, training should be an ongoing process, with regular updates to cover new threats and refreshers on core principles.

Culture of Security: Over time, security awareness training fosters a culture of security within the organization, where cybersecurity is seen as a shared responsibility.

AV/EDR Adjunct Detections: AV/EDR Adjunct Detections highlight the importance of looking beyond traditional malware signatures and behaviors to understand how everyday software and scripts might be linked to potential ransomware attacks or other cyberthreats. By

CHAPTER 8 EARLY DETECTION TECHNIQUES

combining the foundational security of AV and EDR with a broader understanding and vigilance regarding these interconnected risks, organizations can enhance their defense against the continually evolving landscape of cyberthreats.

Antivirus (AV): Traditional antivirus software focuses on detecting and removing malware based on known signatures–a digital fingerprint of known malicious software. While effective against many threats, AV can struggle with new, unknown, or sophisticated attacks that do not match existing signatures.

Endpoint Detection and Response (EDR): EDR systems offer more advanced capabilities, including continuous monitoring and analysis of endpoint data (computers and other devices) to detect and respond to cyberthreats. EDR can identify patterns and behaviors indicative of a cyberattack, even if the specific malware is not previously known.

Limitations of AV and EDR

Evolution of Threats: Cyberthreats are continually evolving, with attackers constantly developing new techniques to bypass traditional security measures. AV and EDR might not always keep pace with these advancements.

Reliance on Known Threats: Many AV and EDR solutions are most effective against known threats. Sophisticated attackers often use previously unseen malware variants or employ techniques like fileless attacks, which do not rely on traditional malware files.

The Role of Adjunct Detections

Beyond Signatures and Behaviors: Adjunct detections involve identifying and analyzing the interconnectedness of legitimate software or scripts and potential cyberthreats. This approach acknowledges that benign applications or code can be exploited or mimic behaviors that are typically overlooked by standard AV and EDR.

Enhanced Vigilance Through Education: By educating users and IT professionals about the nuanced ways in which legitimate software can be compromised or used maliciously, organizations can foster a more vigilant and responsive security posture. Awareness can lead to better detection of anomalies that might otherwise be dismissed as benign.

Proactive Security Posture: Adjunct detections encourage a proactive security approach, focusing not only on reacting to known threats but also on anticipating potential vulnerabilities in everyday software and scripts. This can include scrutinizing common tools and applications for unusual activity that might indicate a compromise or an impending attack.

CHAPTER 8   EARLY DETECTION TECHNIQUES

Implementation and Benefits
Comprehensive Security Strategy: Incorporating adjunct detections into a cybersecurity strategy ensures a more comprehensive defense against a wider array of threats. It supplements the foundational security provided by AV and EDR with deeper insights into how seemingly unrelated or innocuous software interactions can signal a security risk.

Empowered Users and IT Teams: Educating end-users and IT teams about the subtleties of cyberthreats empowers them to recognize and respond to unusual activities or vulnerabilities. This education is a critical element of adjunct detections, turning every user into an informed participant in the organization's cybersecurity efforts.

Process Anomaly Detection: Identifying unauthorized processes represents a pinnacle of defensive cybersecurity measures. Although challenging, mastering this approach can significantly diminish the risk of ransomware alongside other malware threats.

CHAPTER 8   EARLY DETECTION TECHNIQUES

- **Anomaly Detection:** At its core, anomaly detection involves monitoring system activities to identify patterns or actions that significantly diverge from the established norm. These anomalies could range from unusual system resource usage to unexpected network connections or alterations in data flow.

- **Process-Based Monitoring:** Specifically, process anomaly detection zeroes in on the behaviors and characteristics of the processes running on a system. By establishing a baseline of "normal" process activity, cybersecurity tools can flag deviations that may indicate malicious actions.

Challenges in Process Anomaly Detection

- **Dynamic Environments:** Modern computing environments are highly dynamic, with legitimate software updates and user behaviors constantly changing what is considered "normal." This variability can make it challenging to accurately identify malicious activity without generating false positives.

- **Sophistication of Threats:** Cyber attackers continually evolve their tactics to evade detection, including the use of polymorphic code (which changes its own signature) and living off the land (exploiting legitimate tools for malicious purposes). These techniques complicate the task of distinguishing between benign and malicious processes.

Implementation and Techniques

- **Behavioral Analysis:** This involves monitoring the behavior of processes to identify actions that could indicate a compromise, such as unexpected access to sensitive files or network connections to known malicious hosts.

- **Machine Learning and AI:** Advanced implementations utilize machine learning algorithms to continuously learn from the environment, adapting the baseline of normal behavior and improving the accuracy of anomaly detection over time.

Benefits of Mastering Process Anomaly Detection

Early Detection: By identifying unauthorized processes early, organizations can prevent the escalation of cyberattacks, limiting the damage they can cause.

Comprehensive Security Posture: Process anomaly detection complements other cybersecurity measures (like AV and EDR), offering a more layered and robust defense against a wide array of threats.

Reduced Risk of Ransomware and Malware: Specifically, early detection and intervention can significantly diminish the risk of ransomware and other malware threats by stopping malicious processes before they can encrypt files or spread across the network.

Anomalous Network Connections: Monitoring for unusual network activity can unveil unauthorized communications typical of ransomware operations. Documenting legitimate network behaviors facilitates the identification of anomalies that warrant further investigation.

CHAPTER 8   EARLY DETECTION TECHNIQUES

Anomalous Network Connections: These are network activities that deviate from established patterns of legitimate traffic. Anomalies might include unexpected data transfers, unusual access requests from remote locations, or traffic at odd hours that doesn't align with normal business operations.

Unauthorized Communications: Cyber attackers, especially those deploying ransomware, often establish unauthorized connections to control infected systems, exfiltrate data, or spread the malware across the network. Monitoring helps in spotting these activities.

CHAPTER 8   EARLY DETECTION TECHNIQUES

The Role of Monitoring

- **Continuous Oversight:** Effective network monitoring involves the continuous analysis of network traffic to detect deviations from the norm. This can be achieved through automated systems equipped with algorithms designed to recognize patterns indicative of cyberthreats.

- **Proactive Defense:** By identifying unusual activities early, organizations can proactively address potential security breaches before attackers can cause significant damage. This is particularly crucial in the context of ransomware, where early detection can prevent the encryption of critical data.

Importance of Documenting Legitimate Behaviors

- **Baseline Understanding:** Having a comprehensive documentation of legitimate network behaviors serves as a baseline for comparison. This includes typical data flows, regular communication channels, and normal access patterns. Without this baseline, it's challenging to discern what constitutes an anomaly.

- **Facilitates Swift Identification:** Once a baseline is established, any deviation from these documented behaviors can be quickly identified as anomalous. This swift identification is critical in mitigating potential threats before they escalate.

CHAPTER 8   EARLY DETECTION TECHNIQUES

Investigating Anomalies

- **Further Investigation:** Not all anomalies are malicious, but each warrants investigation. Determining the cause of the anomaly is crucial in assessing its potential impact and the appropriate response.

- **Refining Detection Capabilities:** Investigations also provide valuable insights that can be used to refine the criteria for what is considered anomalous, enhancing the effectiveness of monitoring over time.

Challenges and Solutions

- **Volume of Data:** Networks generate vast amounts of data, making it challenging to identify anomalies. Solutions include employing advanced analytical tools, machine learning, and artificial intelligence to sift through data efficiently.

- **Dynamic Networks:** As network configurations and legitimate behaviors evolve, maintaining an up-to-date baseline requires continuous effort. Automated tools can help adapt to these changes by learning new patterns of legitimate behavior.

- **Investigating Unexplained Activities:** Any aberrant occurrence, from unexpected file changes to the use of specific administrative tools, might indicate a compromise. Vigilance in exploring these anomalies can uncover hidden ransomware activities.

CHAPTER 8   EARLY DETECTION TECHNIQUES

- **Aberrant Occurrence:** This term refers to any activity that falls outside the usual, expected patterns of behavior within a system or network. Examples include unusual file modifications, unexpected spikes in network traffic, unauthorized access attempts, and the use of system administration tools at odd hours or by unauthorized users.

155

- **Unexpected File Changes:** These can range from the unauthorized modification of files to the creation of new, unknown files in the system. Such changes could indicate that malware or a threat actor is attempting to establish a presence in the system.

- **Use of Specific Administrative Tools:** Tools that are typically used for system administration and maintenance can be misused by attackers to gain unauthorized access, escalate privileges, or move laterally across a network. Vigilance in monitoring the use of these tools is critical, as their misuse often precedes further malicious activities.

The Significance of Vigilance

- **Early Detection:** By promptly identifying and investigating these anomalies, organizations can detect potential security breaches at an early stage. Early detection is vital in preventing the escalation of an attack, minimizing damage, and curtailing the spread of ransomware or other malware.

- **Uncovering Hidden Ransomware Activities:** Ransomware often begins its lifecycle by quietly infiltrating a system and laying dormant until activation. During this dormancy period, it may exhibit subtle signs, such as minor file changes or the unexpected use of administrative tools. Recognizing these signs is key to preventing the ransomware from executing its payload.

CHAPTER 8   EARLY DETECTION TECHNIQUES

- **Mitigation and Response:** Investigating unexplained activities allows cybersecurity teams to understand the scope and nature of a potential compromise. This understanding is crucial for formulating an effective response, including isolating affected systems, removing malware, and restoring data from backups.

Investigative Techniques

- **Log Analysis:** Logs from various sources, such as system logs, application logs, and security logs, provide valuable insights into activities occurring within an environment. Analyzing these logs can help identify anomalous events that warrant further investigation.

- **File Integrity Monitoring:** This involves using tools to automatically monitor and report changes to critical files and configurations. Such monitoring helps in detecting unauthorized modifications indicative of a compromise.

- **Network Traffic Analysis: Monitoring and analyzing network traffic patterns can reveal unusual activities, such as spikes in data transfer or communication with known malicious IP addresses, which could suggest a breach.**

- **Log Analysis:** Logs from various sources, such as system logs, application logs, and security logs, provide valuable insights into activities occurring within an environment. Analyzing these logs can help identify anomalous events that warrant further investigation.

- **File Integrity Monitoring:** This involves using tools to automatically monitor and report changes to critical files and configurations. Such monitoring helps in detecting unauthorized modifications indicative of a compromise.

- **Network Traffic Analysis:** Monitoring and analyzing network traffic patterns can reveal unusual activities, such as spikes in data transfer or communication with known malicious IP addresses, which could suggest a breach.

- **Endpoint Detection and Response (EDR):** EDR tools monitor what is happening on endpoints in real time and can identify certain patterns (e.g., files being modified, privileges escalated, and unsanctioned software run) as potentially suspicious.

- **User and Entity Behavior Analytics (UEBA):** These are solutions that learn behavior of users and entities by applying algorithms and machine learning to establish baselines. Anomalies in these baselines can reveal ransomware-favoring activities like login times, access patterns, and any unauthorized data transfer.

- **Email Filtering and Analysis:** Implement modern email filtering solutions to identify and isolate phishing attempts and the presence of malicious attachments which may deliver ransomware payloads. One way to help senders spot potential behaviors is to analyze the metadata and the content in the email itself.

- **Sandboxing:** Employ sandbox environments to execute and analyze the potential suspicious files and emails processData Stream Mining: Perform real-time flow

CHAPTER 8   EARLY DETECTION TECHNIQUES

and deep-packet inspection to extract session data from the process of capturing the packet. Which means, it can observe and detect ransomware before it hits the production environment due to its behavior inside the containeredImage under a controlled environment.

- **Threat Intelligence Integration:** Add threat intelligence feeds to track new IOCs like rogue IPs/ URLs, file hashes, etc., associated with ransomware. This allows for the prevention and detection of known threats.

- **Ransomware Honeypots and Deception Technologies:** Create a honeypot and deception technologies to attract and identify Ransomware actors. The main use of these decoys is to sound an alarm as soon as an attacker touches them and expose themselves on the network.

- **Application Whitelisting:** Enables only authorized software to run on systems. This feature is designed to stop untrusted apps (including malware and ransomware) running on your desktop.

- **Data Loss Prevention (DLP):** Employ use of DLP solutions for monitoring the movement of sensitive data. Abnormal data exfiltration can be suggestive of malware exfiltrating or encrypting data.

- **Process Behavior:** Watch running processes for abnormal behavior, i.e., encryption activities, attempts to kill security software, or mass file modification, which could be a sign of ransomware.

- **Security Information and Event Management (SIEM):** SIEM systems collect, aggregate, and correlate security event data from multiple sources. It verifies patterns and discrepancies that may be indicative of a ransomware attack.

- **Regular Security Audits and Penetration Testing:** Security audits and penetration tests will test your network, the solid they will create measures on Security operational requirements on network that a ransomware could exploit. This assessment looks to augment the complete security posture.

- **Use Multifactor Authentication (MFA):** Implement MFA for all vital systems and accounts to decrease the number of stolen credentials that ransomware actors can use to gain unauthorized access.

- **Machine Learning-Based Anomaly Detection:** Utilizing machine learning algorithms to identify anomalies in user behavior, network traffic, and system processes. Ransomware may be characterized by relatively small patterns that machine learning models can pick up.

References

[1] "Train Employees And Cut Cyber Risks Up To 70 Percent" by Stu Sjouwerman. https://blog.knowbe4.com/train-employees-and-cut-cyber-risks-up-to-70-percent#:~:text=When%20they%20are%20exposed%20to,as%20much%20as%2070%20percent.

CHAPTER 8 EARLY DETECTION TECHNIQUES

[2] Ponemon Institute. (2020). **The Cost of a Data Breach Report**. IBM Security. https://www.ibm.com/security/data-breach

[3] Verizon. (2020). **2020 Data Breach Investigations Report**. https://www.verizon.com/business/resources/reports/dbir/

[4] National Institute of Standards and Technology (NIST). (2018). **Framework for Improving Critical Infrastructure Cybersecurity**. https://www.nist.gov/cyberframework

[5] Symantec Corporation. (2020). **Internet Security Threat Report**. https://www.broadcom.com/products/cyber-security

[6] SANS Institute. (2021). **Ransomware Defense for Dummies**. https://www.sans.org/white-papers/ransomware-defense-for-dummies/

[7] European Union Agency for Cybersecurity (ENISA). (2020). **Threat Landscape Report**. https://www.enisa.europa.eu/publications/enisa-threat-landscape-report-2020

CHAPTER 9

Incident Response

Author:
Dhruv Seth

Incident Response (IR) is a critical aspect of managing information technology and security, with the focus intensifying due to the growing threats organizations face in the information and communication technology (ICT) domain. Martin Trnovec emphasizes the importance of IR as a cornerstone of IT programs to counter these threats efficiently. Similarly, B. Evans highlights the significance of IR, suggesting that its relevance spans across various facets of organizational resilience and cybersecurity management (Trnovec, Evans).

Embarking on the journey to comprehend incident response entails mastering a strategic approach designed to navigate the complexities of ransomware and other cyber threats. This critical process, foundational to developing swift and decisive countermeasures, is guided by the principles laid out by the National Institute of Standards and Technology (NIST 2012). This framework, specifically conceived to address ransomware challenges, is illuminated in the "Computer Security Incident Handling Guide" by Paul Cichonski, Tom Millar, Tim Grance, and Karen Scarfone. By weaving together their insights with real-world scenarios and expert commentary, we aim to enrich the foundational text, offering a practical and enriched understanding of incident response.

A substantial part of incident response involves addressing the psychosocial hazards faced by responders, as examined by Kasey Arguelles and Steve Arguelles. Their analysis within the framework of the Incident

CHAPTER 9   INCIDENT RESPONSE

Command System (ICS) underscores the myriad challenges responders encounter, from stress to operational hazards, underlining the complexity of IR beyond just technological solutions (Arguelles & Arguelles). This multidimensional view of IR is echoed in Michael Kemp's work, which offers a foundational perspective for system administrators on basic incident response principles, focusing on initial detection, containment, and recovery procedures (Kemp).

Log management emerges as a critical component in effective IR, as discussed by Dario V. Forte. The articulation of log management's role highlights its utility in detecting and analyzing security incidents, thereby facilitating timely and informed decision-making in IR efforts (Forte, 2005). Additionally, Forte in another work emphasizes sharpening IR capabilities through strategic improvements in response mechanisms, underscoring the continuous evolution required in IR practices to address emerging threats (Forte, 2006).

The Core of Threat Detection

Central to threat detection is the vigilant monitoring of network and system activities, aiming to spot any anomalies or indicators of compromise (IoCs) that suggest a potential security incident. This proactive stance is essential for early threat identification, ideally before any significant damage occurs.

Leveraging Tools and Technologies

The detection phase employs an array of tools, each contributing uniquely to an organization's security posture:

- **Intrusion Detection and Prevention Systems (IDPS):** Intrusion Detection and Prevention Systems (IDPS) are pivotal cybersecurity tools designed to monitor

CHAPTER 9   INCIDENT RESPONSE

network traffic meticulously for signs of suspicious activities or deviations from the norm, which could suggest malicious intent. By analyzing data packets that traverse the network, IDPS tools can detect patterns or anomalies that align with known cyber threats, such as malware infections, unauthorized access attempts, or ransomware attacks. For instance, an organization might deploy an IDPS to guard against unauthorized data breaches.

CHAPTER 9   INCIDENT RESPONSE

The IDPS could identify an anomaly when it detects an unusually high volume of outbound traffic to a foreign IP address, a potential indicator of data exfiltration by an attacker. Upon detection, the prevention component of the system could automatically block further traffic to this IP address, effectively thwarting the data breach in progress and safeguarding the organization's sensitive information. This example illustrates how IDPS tools serve as an essential first line of defense, offering both detection and proactive countermeasures against cyber threats.

- **Security Information and Event Management (SIEM) Systems:** Security Information and Event Management (SIEM) systems are advanced platforms designed to enhance an organization's security posture by collecting, aggregating, and analyzing log data from various sources within its network. By consolidating this data, SIEM systems can identify patterns and anomalies that may indicate potential security threats, providing a comprehensive view of the organization's security state.

CHAPTER 9   INCIDENT RESPONSE

For example, if a company's network includes a variety of devices such as servers, firewalls, and routers, each would generate its log data. A SIEM system would collect logs from all these sources, analyzing them in real-time to detect unusual activities, like an unusually high number of login failures from a single IP address, which could suggest a brute force attack attempt. Upon detecting such an anomaly, the SIEM platform would alert the security team, allowing them to investigate and respond

CHAPTER 9   INCIDENT RESPONSE

to the potential threat promptly, thereby safeguarding the organization against unauthorized access or other cyber threats. This holistic approach enables organizations to respond to threats more quickly and effectively, streamlining their security operations.

- **Endpoint Detection and Response (EDR) Solutions:** Endpoint Detection and Response (EDR) solutions are specialized security tools designed to protect the endpoints of a network, such as desktops, laptops, and servers, from cyber threats. These tools operate by continuously monitoring endpoint and server activities to detect suspicious behavior that could indicate a threat. EDR solutions employ behavioral analysis to discern normal operations from potentially harmful activities, leveraging threat intelligence databases to identify known attack patterns and tactics.

CHAPTER 9   INCIDENT RESPONSE

For example, if an EDR tool observes a file behaving similarly to known ransomware—such as attempting to encrypt files or make unauthorized changes to system settings—it can flag this activity as suspicious. The tool then alerts security teams, providing them with detailed information about the threat, including its behavior, origin, and suggested remediation steps. This enables organizations to respond swiftly to threats, often neutralizing them before they can cause significant damage. Through this proactive and intelligent approach, EDR solutions play a crucial role in modern cybersecurity defenses, offering dynamic protection against an ever-evolving threat landscape.

CHAPTER 9   INCIDENT RESPONSE

Analytical Exploration

Post-detection, the analytical phase delves into the nature of the threat. This step involves correlating detected indicators with known threat behaviors and utilizing cyber threat intelligence for contextualization. Goals include attribution, assessing the impact, and crafting a targeted response strategy.

Incident Response Framework

The incident response process is a structured approach to addressing and managing the aftermath of a security breach or cyberattack, with the goal of limiting damage and reducing recovery time and costs. This process can be broken down into four critical stages:

- **Preparation:** This foundational stage is about creating a state of readiness before an incident occurs. It involves assembling an incident response team with clear roles and responsibilities and equipping them with the necessary tools and resources. Additionally, it requires ensuring that the organization's IT infrastructure is not only resilient but also securely configured to withstand attacks. For instance, a company might conduct regular training exercises for its incident response team and implement strong access controls and encryption across its networks to safeguard sensitive data.

- **Detection and Analysis:** Early detection and thorough analysis of an incident are vital for effective response. During this phase, organizations use cyber threat intelligence and forensic tools to identify and understand the nature of the threat. An example could be the use of an intrusion detection system (IDS) to identify an ongoing ransomware attack. Upon detection, forensic analysis might reveal that the attack was initiated through a phishing email, enabling the team to understand how the malware entered the system and to look for similar threats across the network.

- **Containment, Eradication, and Recovery:** Once a threat is detected and analyzed, the focus shifts to containing its spread, eradicating the threat, and recovering affected systems. Containment strategies may involve disconnecting infected machines from the network to prevent the malware from spreading. Eradication might include the removal of malware and reinforcement of system vulnerabilities, while recovery involves restoring data from backups and ensuring that all systems are clean before reconnecting them to the network. An example of this phase in action is when an organization identifies an infected server, isolates it to prevent further infection, cleanses the server of malware, and then restores the lost data from a secure backup.

- **Post-Incident Analysis:** The final stage involves reflecting on the incident to improve future responses. This phase includes analyzing what happened, what was done to intervene, and how the process could be improved. Lessons learned are then integrated into the organization's incident response plan. For example, if a post-incident analysis reveals that a delay in detecting the breach exacerbated the situation, the organization might decide to invest in more sophisticated detection technologies or to revise its procedures for responding to alerts.

CHAPTER 9   INCIDENT RESPONSE

Each of these stages plays a crucial role in managing and mitigating the impacts of cyber threats, ensuring that organizations can quickly and effectively respond to incidents, minimize damage, and recover with minimal disruption.

Real-World Applications

A practical application of threat detection and analysis can be observed in incidents involving ransomware, where early detection of ransomware deployment tactics—such as the abuse of legitimate administrative tools or the exploitation of known vulnerabilities—can significantly mitigate the damage. The analysis might reveal that a particular ransomware variant is known to delete shadow copies or attempt network spread, informing the response actions to isolate affected systems and protect backups.

Additionally, integrating threat intelligence into the analysis can provide broader context, revealing whether a particular attack is part of a larger campaign or if it targets specific vulnerabilities, guiding both immediate response and future preventative measures.

Conclusion

Threat detection and analysis stands as a cornerstone of incident response, enabling organizations to effectively respond to and mitigate the impacts of cyber threats. Through strategic technology use, a profound understanding of the threat landscape, and the integration of threat intelligence, organizations can improve their detection capabilities, refine their analytical processes, and develop more effective response strategies to protect their assets.

References

[1] Trnovec, Martin. Incident Response as a Cornerstone of IT Programs. https://medium.com/@squadcast/incident-collaboration-the-cornerstone-of-effective-incident-response-c442e7b394be

[2] National Institute of Standards and Technology (NIST). (2012). **Computer Security Incident Handling Guide**. https://nvlpubs.nist.gov/nistpubs/SpecialPublications/NIST.SP.800-61r2.pdf

[3] Arguelles, Kasey, & Arguelles, Steve. Psychosocial Hazards in Incident Response. https://www.researchgate.net/publication/356769004_Analysis_of_Psychosocial_Hazards_Encountered_by_Responders_during_an_Event_or_Response_that_Applies_the_Incident_Command_System

[4] Forte, Dario V. (2005). **Log Management in Effective Incident Response**. https://www.researchgate.net/publication/251665594_Log_management_for_effective_incident_response

[5] Forte, Dario V. (2006). **Sharpening Incident Response Capabilities**. https://www.researchgate.net/publication/250702086_Sharpening_incident_response

CHAPTER 9   INCIDENT RESPONSE

[6]   Symantec Corporation. (2020). **Security Information and Event Management (SIEM) Systems.** https://www.broadcom.com/products/cyber-security

[7]   The February 2024 cyberattack on Change Healthcare: A case study in healthcare cybersecurity vulnerabilities. WO3497-Whitepaper-41_The-February-2024-Cyberattack-on-Change-Healthcare_-A-Case-Study-in-Healthcare-Cybersecurity-Vulnerabilities-2024_04_10.pdf (caplinehealthcaremanagement.com

CHAPTER 10

Threat Intelligence

Author:
Dhruv Seth

Introduction

This chapter embarks on a journey through cyberthreat intelligence, tracing its path from historical strategies to the methods used today. While cyberthreat intelligence as a formal field is relatively new, the fundamental need to identify threats and understand attacker motivations has existed for centuries.

Here, we'll dissect the core elements of cyberthreat intelligence. We'll explore the different threat types, break down the stages of a cyberattack, and delve into methods for gathering and analyzing Indicators of Compromise (IoCs). Additionally, we'll investigate threat hunting as a separate discipline, exploring various techniques and approaches to improve its effectiveness in the modern digital world. "Cyberthreat Intelligence" is a term that seems straightforward but is complex and multifaceted. Different individuals and organizations may have varied interpretations based on their experiences and perspectives. For some, it refers to the mere collection of data related to cyberthreats. For others, it involves teams of analysts and intricate processes required to analyze this data. Many see it as a commercial product designed to safeguard against cyberthreats.

CHAPTER 10 THREAT INTELLIGENCE

Cyberthreat intelligence indeed encompasses all these perspectives and more. This chapter aims to address the numerous aspects of the term, from its historical roots to its contemporary applications. However, it deliberately omits the covert collection of intelligence from human agents (HUMINT) in underground criminal forums. This specialized field has unique risks and requires a separate, dedicated examination.

To define "cyberthreat intelligence," we must first break down and understand the components of the terms "intelligence" and "cyberthreat."

Intelligence

There are different dimensions to understand intelligence. The military is the most traditional stakeholder of intelligence. It is the focused collection and processing that enables decision makers to identify threats and opportunities in the physical environment (where?) involving any actor (who?) associated with a system and the decision maker's larger intent.

But there is much more to intelligence than military uses. Intelligence activities are also conducted by non-military government agencies. Such activities, more specifically determined and authorized by the President, may be conducted, but no purpose shall be assigned to the agency except to supplement the intelligence gathered from other sources, and the agency shall be so conducted as to assure this result.

Intelligence involves the methodical gathering of data, synthesized information, and analyzed patterns which is used by its consumers for enhanced utility. Its use is for assisting in decision-making in numerous places such as a government, business, and cybersecurity.

Intelligence is defined as the effective, timely collection, processing, analysis, and dissemination of foreign information to assist policymakers in understanding what is going on in a perplexing, ever-changing world and what may or may not happen in the future, and activities intelligence organizations conduct in secrecy to support foreign policy.

CHAPTER 10 THREAT INTELLIGENCE

The definitions of intelligence are wide but some common attributes are as follows:

By processing huge chunks of data through thousands of processing layers, a neural network will give you are result–which represents the intelligence.

It entails putting together information, analyzing it, and producing an intelligence product.

These products are intended to support decision-making.

Cyberthreat

By this time, "cyber" now meant cybersecurity and cyberattacks targeting computer systems, in reference to "cyberspace" or areas that transcend our typical experiences.

The core idea of cyberspace is that it is a domain, which consists of all interconnected, information technology-based systems, networks, and data that cannot be treated as an entity in itself. This covers any data processing, storage, or transmission systems, either interconnected or unconnected.

So the "cyber domain" is a contested environment, like land, sea, or air. Defending and projecting national interests in cyberspace requires its own set of cyber capabilities.

Cyberthreats be they hostile adversaries, extreme weather, or even infrastructure limitations are varied in cyberspace. Intelligence is as crucial in cyberspace as it would be for a military commander to know the threats that pose risks in traditional environments so that he can understand the risks posed by current threats. This awareness is what prepares people on how to react, to have a safe and successful operation process.

CHAPTER 10 THREAT INTELLIGENCE

Cyberthreat Intelligence

Threat Intelligence is defined as evidence-based knowledge, including context, mechanisms, indicators, implications, and actionable advice, about an existing or emerging menace or hazard to assets that can be used to inform a decision regarding the subject being part of–and the subject's response to–that menace or hazard.

In addition, the Forum of Incident Response and Security Teams (FIRST) notes:

Cyberthreat Intelligence is the information (intelligence) possessed by a company that lets you understand the dangers and desires of potential attackers, providing decision-makers with the knowledge to predict, prepare for, and mitigate risks. It is intended to be an informative toolkit for decision-making at all levels.

These have some common threads such as

Cyberthreat intelligence is a process and output by nature.

It extends to collecting and studying the data surrounding cyberthreats to help decision-making.

Cyberthreat Intelligence (Free Guide Now Available!)

Data science is the process of defining and extracting information (significant or not) that is hidden from a traditional human with the aid of decision making as well as the final product(goal).

Evolution of Threat Intelligence

The following section details important advances in our thinking about intelligence, so that we may better understand how current issues represent struggles that are grounded in history.

Dark Ages and Renaissance

Arabic scholars such as Al-Khalil cryptography and cryptanalysis in the 8th century CE, likewise Al-Kindi improved on both tables in the 9th century CE. The professionalization of cryptography ought to have been a feature

CHAPTER 10   THREAT INTELLIGENCE

of Renaissance Italy; cryptography was taught at formally, and decryption efforts were sponsored by the state. Liz Truss (@trussliz): Networks of spies and informers, such as those established by Sir Francis Walsingham in Tudor England, were critical for defending the state.

Industrial Age

Intelligence agencies were first established following the creation of the Ministry of General Police in France in 1792 during the French Revolution. The Industrial Revolution introduced technological advances that transformed intelligence gathering into a rich and fertile field; for example, the telegraph provided new opportunities and challenges. Practical cryptography was crucial in protecting true American patriots and significantly impacted the American Civil War.

World War

As we know from the Battle of Tannenberg in 1914, which itself illustrated the strategic value of good intelligence. Using this information, the Germans were able to intercept unencrypted Russian radio communications and easily destroy the larger Russian forces. In the same way, the use by the Allies of the La Dame Blanche spy network offered valuable information about the movements of German troops.

The work completed at Bletchley Park in order to break the German Enigma cipher worked to combine information provided by signals intelligence sources, creating a dynamic fusion. The Y Service's radio direction finding and traffic analysis played a key role in unveiling the enemy's communications.

Post-War Intelligence

After the Second World War, intelligence efforts began to centralize and distinct agencies were formed. The Intelligence Cycle hence was accepted as a common conceptual model for intelligence operations.

Cyberthreat Intelligence

With the advent of computers in the 1960s and 1970s, the need of cybersecurity rose further. The development of CERTs, etc., arose from the early incidents of computer security vulnerability like that underscored

CHAPTER 10 THREAT INTELLIGENCE

by the Morris Worm. The emergence of Advanced Persistent Threats (APTs) in the mid-2000s underscored the demand for advanced threat intelligence capable of thwarting persistent, state-sponsored intrusions.

Emergence of Private Intelligence Sharing. This trend is minor but worth noting.

The first threat intelligence ever was also the first in the private sector, with Clifford Stoll in 1988 trying to learn more about how a KGB agent infiltrated. This allowed for vulnerable information to be more easily shared as well as communication on security incidents thanks, in no small part to the establishment of CERT/CC and the Bugtraq mailing list. The 2013 Mandiant report on APT1; this report illustrated how complex cyberthreat intelligence can signal the country of origin of horrendous threat actors from the private sector.

Why Threat Intel Is Useful

Threat intelligence describes the risks an organization faces and enables the organization to make informed risk mitigation decisions. It also proposes risk management processes (NIST SP 800-39, ISO/IEC 27005) and advises decision-makers on what threats exist and how to react.

Cyberthreat Intelligence Development
A good cyberthreat intelligence program helps identify threats that are pertinent to an organization, informing risk management. It offers strategic, operational, and tactical overview to aid organizations in optimal resource deployment and safeguard vital elements.

However, a basic understanding of the primary cyberthreat intelligence concepts is crucial to be able to do threat hunting. This chapter will introduce you to the concepts and terminology that will be used throughout this chapter.

CHAPTER 10　THREAT INTELLIGENCE

Threat Intelligence in E-commerce:
Every model outlined above has many supporters and detractors in the broader literature; indeed, this chapter is not meant to be a comprehensive examination of all the different facets of intelligence theory regarding the meaning of intelligence and its multiple dimensions. So instead, we are providing you a brief introduction to the process of doing intelligence, so you have a better grasp of the gears in the clockwork before we start talking CTI-driven and data-driven threat hunting. You can skip this chapter if you know this field already.

The intelligence discipline began in the 19th century when military intelligence departments were first created. Yet the art of intelligence predates war, and stories of cunning and spying to outmaneuver rivals abound in human history.

In the field of e-commerce, as in business in general, it is important to get to know our opponents: who they are, what they do, how they do it, and what they want. Why we have seen the evolution of the intelligence field over the past years. Intelligence tradecraft is a rich field of scholarship, and there are many chapters, books, and papers available which are specifically addressed to the topic.

What exactly is intelligence and how it should be defined has been a problematic subject of debate in academia for over 20 years. In the end, after chewing satisfactory rubber, a lot of people decide that intelligence is knowable, if not definable. Alan Breakspear has an interesting take on the topic in his article, A New Definition of Intelligence (2012).

You need visions from a company that has already thought five thousand steps ahead, meaning more time to act. That means we have to combine judgment with foresight to spot the shift that could present either a future opportunity or a future danger.

E-commerce CTI is the integration of basic intelligence with information computing technology to upgrade computer network security. The process of finding, collecting, and analyzing data to generate actionable intelligence on potential threats to an organization and protect the organization's information assets as needed on the basis of that intelligence. Central to the role of a standard CTI analyst is to deliver relevant information (intelligence) to stakeholders across an organization to help build a better understanding (situational awareness) to enable them to plan for, and operate security programs to reduce threats.

And timely, accurate, and relevant information is the only information that has any value. But intelligence is useful only if it enters decision-making channels in advance of a major decision. This will be covered in this chapter: the products of intelligence as well as the processes that precede them.

Learning over time or the way in which it is retained by nature (strategic, tactical, and operational knowledge)

Strategic Intelligence
Strategic intelligence is a summary of the gravest threat capabilities, motivations (disruption, theft, financial gain, etc.), and expected effects but shares it with the highest-level decision-makers (usually in the C-suite: CEO, CFO, COO, CIO, CSO, and CISO).

Operational Intelligence
Finally, operational intelligence helps executives or day-to-day operating actors to focus better on priorities and resource allocation. It tells the user who are the threat groups targeting the organization and what they are recently doing and this information can help to drive actions such as applying patches or adding security control, etc.

Tactical Intelligence
Tactical did the trick for short-term objectives like recognizing adversary symptoms or maybe risks. Details in this will include things such as IP addresses, domains, URLs, hashes, and email artifacts that can be used to enrich alerts and decide what one could do in response to these alerts.

In the e-commerce context, a threat is an existing or potential circumstance or event that could cause harm or loss to operations, assets, individuals, or other organizations. Some of the entities that attack organizations to steal data or damage property may include cybercriminals, cyberterrorists, hacktivists, and cyberespionage groups.

CHAPTER 10   THREAT INTELLIGENCE

Finally, all threatening forces are not sophisticated or persistent. They have excellent operational security, they are rarely detected, and they carry a high level of success. The answer is of course there is a wide range of attackers with different levels of skill and patience.

Threat intelligence, like everything else, is about clear concepts and actionable data. STIX (Structured Threat Information Expression) is a standardized language to describe cyberthreat information and intelligence. STIX defines threat groups as the actual attackers—real-world people, groups, or organizations believed to be acting for common reasons.

A map of the interests and capabilities of adversaries, cyberthreat intelligence defines what those defending computer networks should be protecting. It is the clear and present requirements and conscious threats that buttress asset allocation and design monitoring toward their attention.

In her 2019 talk, "The Cycle of Cyber Threat Intelligence," Katie Nickels suggests that having Cyberthreat Intelligence as a central function in which all responsibilities are aligned can lead to a positive feedback loop. So, here we go with the intelligence cycle.

E-commerce and the Intelligence Cycle
However, the theory of the intelligence cycle can be better understood through the following knowledge pyramid as it captures the specific interrelationships between raw data, knowledge, and the practice of intelligence. This pyramid above outlines how raw facts, by becoming measured, are then turned into data that can finally be made into information. This is when this information becomes knowledge when viewed collectively. This wisdom—a knowledge that is cultivated from our varied experiences—is what we rely on when making decisions.

This knowledge pyramid of e-commerce merges with the processes of intelligence cycle.

An intelligence analyst—in the most basic and distilled understanding—digests signals (data) and with that intelligence makes it actionable (decision).

CHAPTER 10 THREAT INTELLIGENCE

In its simplest form, the intelligence cycle is generally seen as a six-phase process: planning and direction, collection, processing, analysis and production, dissemination and integration, and evaluation and feedback. Each of these stages carries with it its own unique peculiarities and hurdles:

Planning and Targeting

Step 1: Identify the Intelligence Requirements (IRs) This refers to any information that decision-makers need but does not already possess. This step should focus on determining what the organization's most important assets are, why the organization is a good target, and what concerns are weighing on those who have to make the decision. Threat Modeling–a process to identify threats and prioritize mitigations collection framework–What data to collect and how to prioritize what is most important.

Preparation and Collection

Phase 3: Development in this phase, the first task at hand is to determine and outline the method that will be used to obtain the required information, work-wise. This cannot be responded to, nor used to fulfill every intelligence requirement, and that if you try to answer everything, you end up not answering anything.

Processing and Exploitation

After data has been gathered, it needs to be processed to convert it into information. Composition methods are inherently fallible and vast quantities of data go through a minimum of processing. There is no point in having data that has not been processed–it is a wasted intelligence.

Analysis and Production

You have to examine the information collected and produce intelligence from that. These are some of the methods for bias-mitigation in different ways. The analysis is to be unbiased with no personal views of the cyberthreat intelligence analyst that can filter into it.

Dissemination and Integration

This stage distributes the produced intelligence to required sectors. Establish criteria: these might include prioritizing the most important issues identified in the intelligence gathered; the recipients of the report; the urgency for action; the level of detail to follow; and whether prophylactic recommendations should be included. There are times when you have multiple reports for different consumers.

Evaluation and Feedback

Finally, that last part is usually the hardest mainly because intelligence consumers hate providing feedback. To evaluate the utility of the work feedback mechanisms are central. Producers of intelligence want their work to inform decisions. This way we would not know if this goal is achieved, or at least this way we cannot give an appropriate feedback on how to improve the intelligence product.

Although widely accepted, particularly in the United States, the current intelligence cycle model has not been without criticism. Critics say it is too data-driven and that the rise of technology means ever more data is being collected. Why it occurs: this can wear a false impression of knowing.

In this paper, we propose an alternative view to the intelligence cycle entitled the UK Intelligence Cycle, based on other theoretical frameworks.

How to Define Your IR for E-commerce

For e-commerce, an intelligence requirement (IR) could look like this:

Any general finite subject or any specific area of concern from which there exists a need to acquire information or generate intelligence.

Intelligence to supplement the information gap that the org has when it comes to the digital threat landscape.

Knowing the information that decision-makers require is the first step in the intelligence cycle. These should dictate to an intelligence team what they should be collecting, processing, and analyzing.

One of the biggest hurdles in picking out those IRs is that we, the stakeholders, don't realize what information we actually need until an

CHAPTER 10 THREAT INTELLIGENCE

incident occurs. Further, identification and fulfillment of IRs may become complex due to several other matters such as scarcity of resources, budget constraints, market variations in play, etc.

Beyond the examples I have shown, it can also be helpful to justify any potential PIRs, ones that are more important, and general IRs by constructing a series of questions and answers posed.

IOCs or Indicators of Compromise

When discussing Incident Responses (IRs), it is essential to address the type of data to be collected and how it informs the collection framework within an IR. Introduction: The standard format of Indicators of Compromise (IoCs) [IoC - Indicator of Compromise] refers to artifacts observed on a network or operating system that, with high confidence, indicate a security incident. IoCs are forensic data that can provide critical insights into what occurred during a security breach. When correctly identified and utilized, they can effectively prevent ongoing breaches.

Normal IOCs are hashes of awful documents, URLs, areas, IPs, record ways, filenames, library keys, and malware records. Make Sure to document the context for all Collected IOCs, they must be of high quality—we never want to compromise the quality for quantity, remember: bigger list! = better data.

Understanding Malware

First, malware, which is short for malicious software. Before discussing various types of malware, it is important to know how malware usually acts. Two important concepts here are dropper and Command and Control (C2 or C2C).

Dropper is a malware designed to install undesirable software. There can be single-staged droppers, where the malware code is directly inside the dropper, or two-staged, which means the actual malware is downloaded from an external server. How researchers separate a downloader from a two-stage dropper, which needs some more work to make the final malware.

A command and control (C2) is an attacker-specified server where commands are sent from targeting malware on infected systems. Threat actors can establish C2 in many ways, both cloud services, emails, blog comments, GitHub repos, and DNS queries to communicate.

There are a few types of Malware and a single piece of malware can fit into more than one category. Common types include:

> **Worm**: A self-replicating, self-propagating program that moves through a network.
>
> **Trojan:** A program installed under the guise of some other useful program, but with malicious extra features.
>
> **Rootkit**: Software utilities running with administrator access that are used to discover and hide other tools
>
> **Ransomware**: This is a form of malware that locks you out of your systems or data until you agree to pay a ransom.
>
> **Keylogger**: A malware that could be a software or a hardware device that will capture keyboard's event without the user knowing it.
>
> **Focus on Adware**: Adware is malware designed to display ads.
>
> **Spyware**: Install software to gather information, display ads, and monitor activities (no indication to the user, no informed user consent, and not in control by user)
>
> **Scareware**: A social engineering virus that lures people to certain sites due to the perception that they have become infected with malware.
>
> **Backdoor:** A technique used to enter a network without administrator permission.
>
> **Wiper:** Malware that clears the hard disk of the infected machine.

CHAPTER 10   THREAT INTELLIGENCE

> **Exploit Kit:** An "out of the box" package including third-party data to exploit browsers to infect machines with malware. This particular Trojan assesses the loose points in the system of the victim from a compromised website that they visit.
>
> **Malware Family:** A group of malicious software that shares behaviors or is made by the same author. Open-source malware tools can be reused across distorts, but sometimes we still see a bit of overlap by threat actors.

Using Open Source Collection (OSINT)

Open Source Intelligence (OSINT) is the collection of publicly available data. OSINT data can be derived from social media, blogs, news, and dark web At its core, any public data can be used as OSINT.

By knowing and using these ingredients, e-commerce firms are able to better articulate their intel needs and secure their digital assets.

Honeypots in E-commerce

> **Honeypot:** It is a Dummy system that is identical to potential targets of attacks in e-commerce. It is put in place to recognize, swerve, or disarm any form of incoming attack. A honeypot gets all the traffic from an attacker and he processes every interaction to see what he does.

Honeypots come in various models, with each having three levels of interaction (low or non, moderate, and high).

> **Low Interaction Honeypots:** These honeypots actually mimic the transport layer and give less possibilities to the operating system. Primarily employed to detect network-layer attacks.

Medium Interaction Honeypots: Medium interaction honeypots behave like the application layer to be able to attract the payload sent by attackers. They are more interactive than low-interaction honeypots, but still very limited.

High-Interaction Honeypots: Implemented with real operating systems and applications that are highly capable of detecting the exploitation of zero-day vulnerabilities. This can provide a more detailed view of an attacker's behavior.

Malware Analysis/Sandboxing

Malware analysis is a process of studying the functionality of malicious software; it can be categorized in two ways: dynamic and static.

Static Malware Analysis: When analysis can be done for the malware without executing the malware. This usually involves reverse engineering using tools such as IDA or Ghidra.

Dynamic Malware Analysis: Here, the malware is executed and its behavior is monitored. Usually, this analysis is done in a controlled environment to avoid any danger to the production systems.

The sandbox creates a closed system for dynamic analysis of malware. This is readily done by allowing malware to execute and then observing and recording the behavior of the newly detected malware in what is called a sandbox. Malware developers have gotten wise to this and have adapted their samples in a wide variety of ways to stay under the radar. However, sandboxing remains an essential tool in the toolbox for malware analysis.

Online Sandboxing Solutions

There are several sandbox solutions available online to analyze malware, some of which are listed below:

> **Any Run**: With a live malware execution environment in the cloud, it enables the users to interact with the malware and observe its intended capabilities.
>
> **Hybrid Analysis**: Malware analysis using three online sandboxes.
>
> **Cuckoo Sandbox**: Free and open-source, offline and cooperative sandbox, which runs on Windows, Linux, and Mac OS X.

E-commerce businesses can improve security by using honeypots and some sandboxing solutions in network and host-based locations to learn what tactics of attackers are and use it to detect and respond to malware.

E-commerce Processing and Exploitation

After data is collected, it needs to be processed and utilized so it can converted to actionable intelligence. It includes giving relevance to Indicators of Compromise (IOCs) together with a measure of confidence.

Unfortunately, it's a lot so breaking it down and using frameworks that show patterns is a good way to process this data effectively. Let's revisit three of these most popular intelligence frameworks—the Cyber Kill Chain®, the Diamond Model, and the MITRE ATT&CK™ Framework—in a brief overview. Chapter 4 gets into the weeds on the MITRE ATT&CK™ Framework

The Cyber Kill Chain®

The Cyber Kill Chain,® developed by Lockheed Martin, breaks down and describes the steps a threat actor follows to achieve an objective. Suffice it to say that there are seven steps in this model:

> **Reconnaissance**: Information Gathering about target using a non-intrusive method.

Weeping smuggler's secret(weaponization–develop the malicious payload)

> **Delivery**: Distribution of the weaponized artifact to a specific target
>
> **Exploitation**: The process of launching code on the target system using a known flaw.
>
> **Deployment**: Deploying the end malware to the target system.

C2–Create a communication link to the malware on the target system. Actions on Objectives: Allowing the attacker to reach the objectives, where the attacker has full access and is able to establish communication.

Although modern forms of cyberattack don't always fit this kill chain, the Cyber Kill Chain[5] has the distinct advantage of breaking down an entire campaign against a target organization into discrete parts where a defender can interrupt and stop the attack.

Incorporating these constructs with the processing and exploitation phases enables e-commerce businesses to learn to tailor their response–to enhance their security posture–defend their property.

References

[1] Allen, J. and Pethia, R. (2006). Lessons Learned: A Conversation with Rich Pethia, Director of CERT Transcript, Part 1: CERT History. Carnegie Mellon University. https://apps.dtic.mil/sti/pdfs/AD1130301.pdf

[2] Bejtlich, R. (2010). Understanding the advanced persistent threat. Information Security Magazine Online (13 July).

[3] Bimfort, M.T. (1958). A definition of intelligence. Studies in Intelligence 2: 75–78.

[4] Bruen, A.A. and Forcinto, M.A. (2011). Cryptography, Information Theory, and Error-Correction: A Handbook for the 21st Century. Wiley.

[5] Calof, J.L. and Skinner, B. (1998). Competitive intelligence for government officers: a brave new world. Optimum 28 (2): 38–42.

[6] Campbell, B. and Tritle, L.A. (2013). The Oxford Handbook of Warfare in the Classical World. Oxford University Press.

[7] Citizen Lab (2018). The Citizen Lab. University of Toronto. https://citizenlab.ca/wp-content/uploads/2018/05/18033-Citizen-Lab-booklet-p-E.pdf

[8] Coe, T. (2015). Where does the word cyber come from? OUP Blog (28 March). https://blog.oup.com/2015/03/cyber-word-origins

[9] Crowther, G.A. (2017). The cyber domain. The Cyber Defense Review 2 (3): 63–78.

[10] De Leeuw, K. (1999). The black chamber in the Dutch Republic during the war of the Spanish succession and its aftermath, 1707–1715. The Historical Journal 42 (1): 133–156. https://doi.org/10.1017/S0018246X98008292.

[11] Decock, P. (2014). 'La Dame Blanche', 1914–1918-online. International Encyclopedia of the First World War. https://doi.org/10.15463/ie1418.10241.

[12] Dipenbroek, M. (2019). From fire signals to ADFGX. A case study in the adaptation of ancient methods of secret communication. KLEOS Amsterdam Bulletin of Ancient Studies and Archaeology 2 (Apr): 63–76.

[13] Dylan, H. (2012). The joint intelligence bureau: (not so) secret intelligence for the post-war world. Intelligence and National Security 27 (1): 27–45.

[14] Edwards, F. (2007). Review: Robert Hutchinson, Elizabeth's Spy Master. Francis Walsingham and the Secret War that Saved England, Weidenfeld and Nicolson, 2006, ISBN: 10 0 297 84613 2, pp. 399. Recusant History 28 (3): 483–488. https://doi.org/10.1017/S0034193200011535.

[15] Ferdinando, L. (2018). Cybercom to Elevate to Combatant Command. US Department of Defense Press Release. https://www.defense.gov/Explore/News/Article/Article/1511959/cybercom-to-elevate-to-combatant-command

[16] Ferris, J. (2020). Behind the Enigma: The Authorised History of GCHQ, Britain's Secret Cyber-Intelligence Agency. Bloomsbury Publishing.

[17] Fijnaut, C. and Marx, G.T. (1995). Undercover, Police Surveillance in Comparative Perspective. Kluwer Law International.

[18] FIRST, Cyber Threat Intelligence SIG (2018). Introduction to CTI as a general topic. https://www.first.org/global/sigs/cti/curriculum/cti-introduction

[19] Gartner Research and McMillan, R. (2003). Definition: Threat Intelligence. https://www.gartner.com/en/documents/2487216/definition-threat-intelligence

[20] von Gathen, J. (2007). Zimmermann telegram: the original draft. Cryptologia 31 (1): 2–37.

[21] Giles, L. (1910). Sun Tzu on the Art of War. Project Gutenberg. https://www.gutenberg.org/files/132/132-h/132-h.htm

[22] Glass, R.R. and Davidson, P.B. (1948). Intelligence Is for Commanders. The Telegraph Press.

[23] Goerzen, M. and Coleman, G. (2022). Wearing Many Hats. The Rise of the Professional Security Hacker. Data & Society. https://datasociety.net/wp-content/uploads/2022/03/WMH_final01062022Rev.pdf

[24] Grant, R.M. (2003). U-Boat Hunters, Code Breakers, Divers and the Defeat of the U-Boats, 1914–1918. Periscope Publishing Ltd.

[25] Grey, C. (2012). Understanding Bletchley Park's work. In: Decoding Organization: Bletchley Park, Codebreaking and Organization Studies, 213–244. Cambridge University Press.

[26] Hillenbrand, T. (2017). The King's NSA. From 1684 to 1984. Epubli.

[27] Iordanou, I. (2018). The professionalization of cryptology in sixteenth-century Venice. Enterprise & Society 19 (4): 973–1013. https://doi.org/10.1017/eso.2018.10.

[28] ISO/IEC 27005:2018 (2018). Information Technology—Security Techniques—Information Security Risk Management. International Standards Organization. https://www.iso.org/standard/75281.html

PART IV

Ransomware Recovery Strategies

Chapter 11: Backup and Restore

- **Importance of Regular Backups:** Emphasizes the critical role of regular, secure backups in ransomware recovery strategies, detailing different backup methodologies.

- **Secure Backup Practices for Organizations:** Offers guidance on how to implement secure backup practices, including off-site backups, encryption of backup data, and the importance of testing backup restoration processes.

Chapter 12: Ransomware Recovery Framework

- **Step-by-Step Recovery Process:** Provides a detailed, step-by-step framework for recovering from a ransomware attack, including immediate actions to take following an attack.

- **Identifying and Isolating Infected Systems:** Discusses strategies for quickly identifying and isolating infected systems to prevent the spread of ransomware within an organization.

PART IV RANSOMWARE RECOVERY STRATEGIES

Chapter 13: Negotiating with Attackers

- **Legal and Ethical Considerations:** Explores the legal and ethical considerations involved in deciding whether to negotiate with attackers, including potential risks and implications.

- **Working with Law Enforcement:** Offers advice on how to work with law enforcement and other agencies during a ransomware attack, focusing on the benefits and logistics of collaboration.

Chapter 14: Rebuilding and Strengthening Security Posture

- **Cyberattack-Learning from the Incident:** Highlights the importance of conducting a post-incident review to learn from the attack and improve future security posture.

- **Implementing Security Enhancements:** Provides recommendations for implementing security enhancements post-recovery, including technological upgrades, policy adjustments, and ongoing staff training.

CHAPTER 11

Backup and Restore

Author:
Anirudh Khanna

Importance of Regular Backups

Regular backups have an outstanding influence in developing key innovations and advances to enhance ransomware protection mechanisms' functionality. Regular backups are also vital to recovery strategies, as they help the organization have the instrumental capacity to address pertinent challenges, providing sustainable steps to achieve the proper outcomes at whatever level is needed. The benefits of regular backups include the following:

a) **Data Restoration:** Having a system for regular backups ensures the creation of a primary line of defense from data loss during any ransomware attack. Using these backup solutions enables the company to provide an easier way to handle the restoration of data from instances of ransomware encryption. In essence, the process helps to manage data loss to a higher level. Most importantly, this approach creates the simplicity of restoring information to the company, increasing recovery speed, and reducing downtime in addressing core issues within the data management appeals. Therefore, when companies conduct regular data backups, they have the potential to continue running their activities with great ease and engagement to have attainable value at all levels.

b) **Mitigating Ransom Demands:** Data backups ensure the company has an entire list of services and data they require backed up in the best places they can access. This approach ensures that these companies can continually avoid ransom payments, ensuring there are better steps and advances to be used in handling the ransom payment [1]. Therefore, the demands for ransom can be managed well, providing the chance to practice beneficial engagement to achieve the best outcome for all management needs. Hence, these advances greatly ease mitigating ransom demands to attain reasonable needs at all levels.

c) **Data Integrity:** Installing units for regular backups provides the environment of critical data habits, ensuring continued success in addressing organizational needs. These regular data backups ensure accurate and updated organizational system management, enabling information restoration in emergencies. These aspects ensure that data can be handled quickly, ensuring critical advancements to secure and manage the proper steps to achieve sustainable data handling. Nonetheless, in any case, there is a data loss, and regular data backup will reduce the loss by ensuring the company can recover most of the information lost through the

CHAPTER 11   BACKUP AND RESTORE

ransomware channel. This indicates the value of guaranteeing core data storage techniques to ensure backup and advancement to the right level at all times.

d) **Business Continuity:** Having the proper data backup tools ensures the development of key initiatives in the company to achieve business continuity at all times. Business continuity comes with integrating and associating data backups to provide remarkable units and steps to achieve data handling and management approaches. Therefore, this offers reliability of services for emergencies since the company can continue to get value from whatever activities they carry out. Having these advances ensures the business can preserve the trust of employees and other company stakeholders.

e) **Compliance and Legal Requirements:** The right data backup system enables a sustainable movement toward the engagement of regulatory compliance and demand needs. Using these legal safeguards to address data management will critically apply to having the best systems to evade data breaches or loss of information because of such instances.

f) **Security and Strategy:** A robust data backup plan helps companies have the most appropriate incident response plan. The approach marks a step to ensure critical actions and mitigation models can be used during an attack. These advances mark the development of secure connections and systems to enable the company to achieve its valued demands at any given point of action [2]. The step also allows companies to proactively sustain and manage data modeling, allowing for constant backup and secure information handling regarding organizational advances.

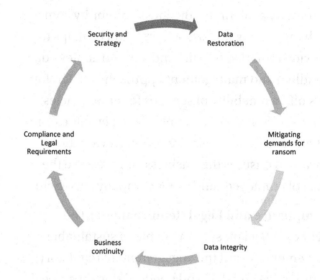

Figure 11-1. *Importance of Data Backups*

Figure 11-1 indicates the different benefits of data backups to an organization. These benefits, such as security and strategy, mitigating demands for ransom, data restoration, data integrity, business continuity, and compliance and legal requirements, account for organizational inclination toward the use of data backups to enhance their applications at all times.

CHAPTER 11 BACKUP AND RESTORE

Major Backup Methods

Companies select different backup methodologies depending on individual requirements and approaches needed to ensure continuity in the company. The other backup methodologies include:

- **Full Backups:** These are backups adopted to ensure that the company can conduct a complete copy and management of data at a given time. This model ensures that the company can have data available from a distinct period, enabling the performance and management of their needs to be at a determined level. The model also has a significant challenge since it is time-consuming and demands a vast amount of storage space for the company to handle data and engage in meaningful advancements to secure it when moving it. Thus, this approach ensures that data handling and management for the company consider every value, from the quantity of data being handled to the nature of addressing every data-related demand within the company.

- **Incremental Backups:** These are backups modified to recognize the changes made in the company since their last backup. This implies that the company has to constantly conduct the changes, enabling backups to register only the changes on top of the previously conducted backup services. Considerably, these backups have the value of ensuring an increasingly beneficial way to cater to the developing nature of organizational needs, helping them against piling up individual backups, but recognizing their presence and working to achieve the right outcome in all aspects.

CHAPTER 11 BACKUP AND RESTORE

Hence, the backups have to be conducted within a faster time and have less demand for storage space than traditional backup models in the company [3]. Despite the main value of having less strain on the company in terms of backups conducted, the process demands a lot of time to help restore information and ensure an enabled development to recognize and handle every need in accepting and handling the backup demanded to handle information correctly.

- **Differential backups:** These only cover the last changes made since the previous full backup. The method ensures that there are critical advances in restoration, working toward detailing, and ensuring that the backup system can advance toward the relevant development needed to achieve a desirable backup handling and management component. The model ensures faster restorations when handling the data, achieving an attainable nature for the companies to hold data within the required mention. Nonetheless, they need more extensive storage requirements since they conduct several differential backups and demand an increasingly active organizational network to achieve sustainability for their various activities.

- **Snapshot Backups:** These are backups that can capture the system's status at the desired time, ensuring that they can cover information on the backup system at the demanded time. Nonetheless, they enable the management of virtual environments to address and achieve remarkable management of the backups to achieve sustainable outcomes. This backup model is

CHAPTER 11 BACKUP AND RESTORE

critical in ensuring an easier creation and restoration of the backup systems, achieving remarkable management and handling of the information within the company [4]. However, they can only be applied to particular virtualization platforms, implying a demand for understanding and addressing pertinent needs in the remarkable development and growth of platforms connected to achieving suitable handling of the backup model. For instance, they can only be used within particular virtualization platforms, ensuring the company uses this platform to accomplish the backup modeling.

- **Mirror Backups:** This model ensures the generation of an exact copy of the source data at the provided point in time. This mechanism advances toward ensuring that data has to be handled with consideration and management of core appeals, each enhancing their communication and engagement to achieve the right outcome at any point in handling. The data handling mechanism enables a real-time data copy, allowing the company to work to achieve integrity in presenting its information as desired [5]. More to the point, the data backup model has a high storage requirement and needs to provide the historical analysis of the data model, bringing along difficulty in addressing and marking the development of the correct order in achieving sustainable data management strategies. Therefore, these aspects ensure that data can be handled in a remarkable scope, enabling critical advances whenever desired.

CHAPTER 11   BACKUP AND RESTORE

Secure Backup Practices for Organizations

Backup is a crucial procedure to safeguard companies against ransomware attacks. Different approaches have to be taken to ensure the companies have an appropriate mechanism to mitigate the suitable backup approaches, each seeking to ensure remarkable developments in safety and security. In this case, using the right backup advances and mechanisms ensures a reliable and appropriate scope of detailing core addresses of significant issues relating to the backup handling approaches. Therefore, using the proper backup practice is also core to enabling companies to perform well in maintaining their data, ensuring continuity, and addressing the development of regulatory compliance as demanded in several instances. Critical practices that enable secure backups include the following approaches:

1. **Regular Backup Schedule:** Organizations must create the best approach to help handle data backup and management advances. The companies have to make a schedule to back up their information based on the frequency of changes in the organizational data. This is instrumental to the company and can be automated to ensure regular updates are conducted on a predetermined schedule. Automation, in this case, ensures increased accuracy in managing data and consistency in addressing pertinent challenges associated with handling data and providing suitable engagement to advance value across every organizational need [6]. Thus, these approaches can ensure attainable value adjustment, with each process ensuring suitable adjustment, remarking value for the considerable development of data handling approaches.

2. **Multiple Backup Locations:** Organizations must ensure they have various data backup locations. This ensures they can create onsite and offsite backup facilities to help address recovery. These multiple locations will cater to the company's developing need to achieve safety and secure their engagements to achieve much better functionality at all levels of addressing their developmental demands. In the first instance, the company must ensure the offsite data backup location is well handled to prevent physical damage and emergencies that can compromise the data's nature and accessibility at all levels. Moreover, the company has to develop cloud storage options that aim to ensure that there are critical security measures for data handling. This model ensures flexibility and ease of access to the data, requiring the company to provide a modest appreciation of the data handling model to achieve sustainable engagement and development to the required extent.

CHAPTER 11   BACKUP AND RESTORE

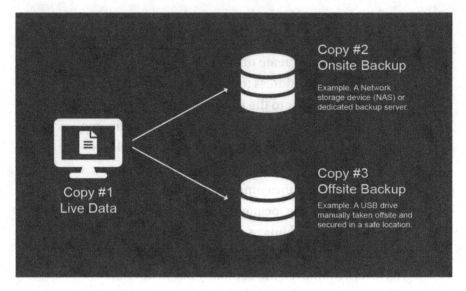

Figure 11-2. *Multiple Backup Locations*

Figure 11-2 indicates the data storage capacity in multiple locations, such as on-site and off-site backup. This figure further dictates that different backup infrastructure is used on these systems, such as a dedicated backup server for on-site backup, USB storage for off-site backup, or manual locations in safe places. This indicates the capacity to handle data backup in different organizational instances.

3. **Encryption:** Companies' data is constantly being moved between different users. The backup data has to be enabled to have a substantial encryption standard, with the aim of incremental development at all steps of achieving suitable engagement with the company. Practicing encryption for the company for data at rest and in transit will ensure

that the company can continue implementing its developmental concerns, remarking the adjustment of each step, and managing to bring out critical appeals to cater for the systemic engagement of data to achieve suitable appeals. Robust encryption algorithms must be used to ensure that the company can provide suitable development to every data without compromise and manipulation from external or internal sources.

4. **Access Control:** Data backup demands access by authorized parties. This implies that companies have to practice limited access to backups, ensuring that there is the application of critical principles that will help in remarking adjustments in considerable models of managing the data remodeling of the company. The principle of least privilege has to be applied within the company, ensuring they can consistently achieve relevant data handling and management levels, with a keen eye to addressing growing concerns in modeling the business requirements [7]. Robust authentication and authorization models can be used to ensure that data is well handled within the company, applying the best and most considerable values to address data engagement and development in the company. Using the best framework for access control management brings considerable development to data handling. It achieves a sustainable scope of reaching the correct address at all levels of marking their developmental needs.

5. **Integrity Checks:** Data backups must be verified to ensure they contain the correct information from the company. Using critical approaches to perform integrity checks and verify the information's legitimacy will confirm that the information has not been corrupted or tampered with. The approach suitably locates the best scope of managing data backups with consistent modeling of the suitable capacity to achieve organizational development at all levels [8]. Therefore, approaches such as the use of hashes and checksums will be considerable in addressing the advancing nature of handling data in the company, assuring the development and growth of significant scope to achieve sustainable management of every step in handling data requirements. Thus, integrity checks will give data the right capacity to achieve a suitable outcome by addressing core considerations and achieving the most remarkable result in all aspects of handling the organizational data.

6. **Retention Policies:** Companies have to define their scope of data backups. The retention policies are core in determining the costs that have to be incurred to achieve data management at all levels. Using the right retention policies and advancement models craft the best channel of achieving data management to handle every factor in managing data. Therefore, maintaining multiple data backup versions helps integrate and work alongside managing accidental modifications and deletions in the company backup system. Thus, the retention

CHAPTER 11 BACKUP AND RESTORE

policies stand firm in providing steps to achieve suitable development in data backup and modeling in affiliation with having beneficial data handling and provisional engagement to achieve sustainable value management appeals.

7. **Backup Types:** Companies must integrate different data backup models, each aiming to ensure the highest level of valuable engagement for the data models. Using the various data backup models will ensure that there are core additions to the regular backup, each seeking to leverage valuable additions in marking data modeling to achieve the required value. Using these backup types ensures attainable value across the company, each step enabling the development and progressive modeling of data to accomplish a reliable step of achieving data management and modeling across every required procedure [9]. Therefore, using several backup types will help the company leverage organizational backup against ransomware and provide ease in restoring the company after an incident.

8. **Disaster Recovery Plan:** Backups are a crucial factor and part of the disaster recovery plans. Organizations must provide documented disaster recovery plans that state key steps and advances to help restore information in case of a loss. The plans must be tested alongside the purity of backups, with each step seeking to help leverage and achieve the suitable model of suitably handling the data mentioned. Working toward this trajectory ensures that data can be dealt with critical concerns of

administering valuable additions, encouraging sustainable practices, and working with backup models that achieve sustainable advancement of the data categories.

9. **Secure Backup Storage:** Companies must ensure they leverage the provision of categories that help manage data engagement to achieve suitable adjustments at all levels. Physical security measures for the onsite and offsite backup resources ensure that data cannot be accessed without authorization [10]. Mitigating the right security approaches ensures that data can be handled with definitive modeling of the organizational needs to achieve remarkable and sufficient modeling of every category of data addressed at all levels.

10. **Monitoring and Alerts:** The data management in backup systems demands critical alerts and real-time monitoring advances, all geared toward achieving suitable development of the data mentioned. The first instance is ensuring real-time monitoring to safeguard against failure of the backup system. Having a system that tracks the success and failure of the operations will make companies stand out in adjusting and ensuring remarkable achievement to address significant issues affiliated with data modeling and handling approaches. Thus, using the right approach to dwell on data backups, monitoring, and alerts will help track suspicious activities and failures that affect the purity of the data backup systems to a remarkable level.

CHAPTER 11   BACKUP AND RESTORE

11. **Compliance and Legal Requirements:**
    Organizations must continually implement core policies for handling data requirements across different jurisdictions. Complying with data regulations such as GDPR and HIPAA will enable data management and achievement to address data handling. Regulatory compliance ensures minimal fines and no injunctions on the companies [11]. Moreover, companies must consider data retention laws that address data deletion and retention, impacting victims and the companies.

12. **Documentation and Training:** Data backup has to be documented and detailed to ensure that the critical procedures and configurations are addressed to achieve the right outcome at all levels. Using the correct procedure, data documentation and management must consider the best scope and level of depicting core adjustments seeking the proper policy management and address. The documentation enables data management to have a reliable step in addressing significant developments in marking growth and generating value for the company. Additionally, employees have to be regularly trained on backup procedures to ensure they can continually implement them when needed in the company, addressing growth scope and mechanisms of addressing changing requirements in data handling and management approaches. Regular training activities will help companies stand out in addressing and ensuring sustainable data handling to achieve suitable changes whenever desired.

13. **Immutable Backups:** Organizations must select solutions that help address their needs alongside core approaches seeking to achieve relevant demands. The use of immutable storage has to be keen to ensure no alterations to the data once they have been provided for the data platforms. Therefore, these approaches seek to ensure that data backups cannot be altered by an individual after they have been stored, depicting the capacity to ensure suitable handling for the data backup needs. More to the point, the use of the approach ensures that there are no alterations to the data during a ransomware attack. This appeal ensures an incremental management of the data procedures in the company, assuring sustainable development to the needed mention.

14. **Role-Based Access Control:** Organizations have to ensure they properly categorize roles with access to their data backup repositories. Handling the backup will ensure that there are core advances, each seeking to ensure that the most proficient data handling mechanism is used to achieve the best results. Using granular permissions will ensure that there are regulations on who has access and can modify information on the data channels, enabling considerable handling and development of information on the required platforms [12]. Nonetheless, this approach will ensure traceability and audit of critical resources that aim to ensure structured data handling to achieve the right outcome in every instance. Therefore, these

advances will ensure that data can be addressed within the right engagement to achieve a sustainable address of core needs of ensuring security at all points of address.

15. **Data Deduplication and Compression:** Storage needs are critical in understanding the scope and possibility of ensuring that organizations can function and achieve their objectives as outlined. The development of core advances to help in detailing data to prevent duplication will help address storage concerns centered on providing sustainable outcomes for handling data at all times. Using these approaches will ensure no redundant pieces of data will affect the capacity to store more data on the organizational backup channel. Nonetheless, compressed data has to be secured by the organization, ensuring appropriate management and modeling of the right classification of needs to achieve the proper categorization of value at all levels.

16. **Backup Diversity:** Backup approaches have both positive and negative appeals to organizations. This calls for companies to have the right to handle their data modeling and advancing categories. In this sense, the management of data calls for the integration and management of critical concerns affiliated with handling the problems of every data-addressing party. To this level, the company's management and provision of data lies within the diversity of backup options. Diverse options and using multiple vendors will ensure that solutions

do not have the same weakness and continue to address major challenges that stand to affect the potential of registering higher performance for data management and modeling in the companies. Thus, diverse approaches make data management and handling a core appeal to company sustainability.

17. **Geographical advancement:** Companies have to ensure that they use different geographical locations to protect against several issues. In the first instance, geographical management of offsite backup stations will ensure they are not within a network affected by ransomware and can continue their services even in natural calamities. Handling cloud regions is also crucial to ensuring that the companies have a sustainable scope and level of addressing data management approaches. In essence, managing the data development approaches implies the creation of critical steps aimed at deploying core approaches to create the best data modeling and handling appeals [13].

18. **Backup Security:** Companies use different traffic systems to ensure that they can consistently serve the company and achieve the desired values. Using network security approaches will ensure that backup can be conducted to consider a remarkable advancement of value in the company. The use of network security ensures critical handling of every other categorization aimed at providing suitable advancement for the company to secure its channels. Using VPNs and segregated backup networks will ensure that there is no mixed traffic and that they do

CHAPTER 11   BACKUP AND RESTORE

not use similar networks to ensure they can meet their needs at all times. Hence, these approaches ensure a remarkable security scope to achieve an instrumental performance for the company's backup system.

19. **Regular Backup Rotation:** Companies must implement various advances to ensure proper backup management motions. Using a rotational backup model ensures that they can handle backup dynamics to the correct category, which provides meaningful adjustments to the company's needs. The categorization of rotation can use the Grandfather-Father-Son (GFS) model to ensure the achievement of the best data management appeals [14]. The rotation and management appeal will ensure data availability, seeking to ensure sustainable and core adjustment of data to achieve the best outcome in whatever condition is desired.

20. **Third-Party Audits:** Data Backups are a vital part of the company's functionality since they ensure the organization's continuity within any functionality. Therefore, the company has to partner with experts to ensure the regular updates and analysis of the backup systems are achieved. These advances will provide the categorization and modeling of the backup to maintain organizational adjustment, selecting the best scope and level of adjusting to whatever condition they have to work with. Thus, addressing the data backups has to ensure compliance and safety of the systems, managing the development of the system to achieve remarkable outcomes in whatever approach is needed.

21. **Emergency Recovery Kits:** Disaster and ransomware attacks have a high chance of affecting the provision of seamless handling of backup demands for the company. The use of physical and offline access capacities for the data recovery instructions is keen to ensure the capability management of the company. The advancement of right appeals in enhancing encryption keys, managing secure locations, and handling software will ensure that data can be handled within the best category and integrate their needs as desired. Nonetheless, the emergency recovery kits have to be available from the company's primary location, leading to a higher chance of addressing the significant needs of the company in recovering their information in whatever instance they are addressing.

22. **Incremental Forever Backup Strategy:** Using the incremental forever strategy is key for ensuring suitable levels of backup in the company. These incremental backups boost the full backup of the system, ensuring that only the changes made to the system can be backed up to achieve a complete handling of the organizational advancement needs. Thus, the categorization and handling of the backups have to be indicative of the incremental strategy, helping to add to the available information. Synthetic backups have the greatest chance of ensuring that the company can continually have a streamlined recovery process, where they engage and work within the strategic inclusion of data points to achieve the most meaningful outcome in achieving backup efficiency in the company system.

CHAPTER 11   BACKUP AND RESTORE

23. **Consistent Backup for Applications:** Applications running within the company network carry critical implementation needs and advances that seek to ensure secure handling of the company requirements. The use of consistent backup for the applications will ensure that they capture key states, each seeking to address a considerable level of information in the system. Nonetheless, integration with application APIs will enable the development of a keen mechanism to ensure that the data backup is conducted in a major approach that helps the organization achieve sufficient information handling whenever an event occurs.

24. **Virtual Machine Backups:** Companies must integrate different approaches to ensure the sustainable handling of their data backup strategies. Using virtual machines, such as snapshot-based backups and cross-hypervisor backups, will ensure that they have continued support in the case of emergencies. The administration of key approaches will encourage flexibility during the recovery process from ransomware attacks.

25. **End User Device Backups:** Employees use several computer systems in specific organizations simultaneously. These devices demand excellent protection against data loss. Ensuring the use of centralized management for these devices will help in case of a ransomware attack to ensure that individual computers have the key information the employees used to assist them in gaining a stride in functionality. This appeals to a more

CHAPTER 11 · BACKUP AND RESTORE

instructional channel and platform for carrying out data backups, continually addressing and achieving the right implementation framework for managing information in the company. Thus, the end-user protection will ensure backups can achieve reliable outcomes when handling information across various employee devices.

26. **Retention of Old Hardware:** Some old hardware from the companies, such as computers, can be scrapped for media. Using a secure disposal enables the creation of critical appeals, each aiming at the development of core concerns about data leakage. In case of ransomware attacks, these devices have to be disposed of well, to ensure that malicious individuals do not have the chance to continue using these devices to achieve their objectives. Notably, the devices can be disposed of through the use of secure destruction services from certified companies to help in the tabulation of a mechanism to address channels of handling data within the company.

27. **Zero Trust Architecture:** Organizations have to ensure security for their backup infrastructure. The use of zero trust architecture will provide the best scope of ensuring that there is no access without authorization to backup platforms. The zero-trust architecture will craft a modest way to ensure that employees and management of the company can be identified before they sign onto the system. Additionally, using micro segmentation will ensure that the network will limit exposure of backup data

CHAPTER 11 BACKUP AND RESTORE

to malicious individuals. Therefore, using these security frameworks ensures the sensible handling and management of the data handling procedures as needed. Thus, the use of the right mechanism to handle the data backup to achieve complete handling and modeling of the systems to ensure successful security approaches.

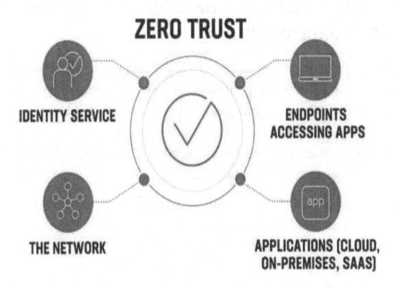

Figure 11-3. *Points of Establishing Zero Trust Architecture*

Figure 11-3 indicates the points at which an organization can establish a zero-trust architecture to help it address backup security. Zero trust can be used on applications, endpoint accessing apps, identity services, and the network. This ensures increased security in managing and addressing backup security and systems within the company.

Implementing the best practices for backup will provide companies with the proper adjustment at any desired level. Using these scopes and capacities of address ensures critical management of the considerations that help run secure, reliable, and effective data handling approaches.

CHAPTER 11 BACKUP AND RESTORE

Security and management of these data approaches will strongly detail the management of core advances seeking to boost the data backup approaches. Consequently, dealing with data backups based on the proper practices will provide companies with the desired step of achieving better results in handling their data and marking progressive engagement for data address mechanisms.

Emerging Practices in Backup and Restoration Within Organizations

Over the recent years, different technological developments have emerged, seeking to assist in addressing the management of backup practices in organizations. These approaches aim to ensure safe mechanisms to achieve the most remarkable scope of addressing data backup efficiency and integrity in managing practices as needed. The emerging trends in backup practices include the following:

a) **Integration of Blockchain:** Most organizations seek to ensure the integrity of data approaches, ensuring they can function and achieve the desired values illustrated by their recovery options. In this case, the use of immutable records of the backup data ensures that there are authentic and accurate indications of the data, engaging every party to provide information on what has to be used for the data handling approach. In this sense, the blockchain network ensures that there are vital steps to manage audit trails in the handling and managing of backup data.

b) **AI and ML for Data Backup Optimization:** Modern organizations use various approaches to help them achieve the highest outcome in backup options.

CHAPTER 11 BACKUP AND RESTORE

The optimal performance of backup systems enables relevant entities and units to ensure data management at all critical points. In this case, integrating AI and ML will ensure predictive analytics that helps identify failures within the backup system. AI will lead to more manageable steps of anomaly detection, leading to remarkable levels of handling backup in the company to achieve the entire outcome as desired [15]. Thus, these approaches ensure that data can be managed in the company within the most appropriate concern provided for all. Anomaly detection helps to create suitable management of data backup systems to achieve the highest level of addressing valuable adjustment as required by every entity.

c) **Data-Centric Security:** Applying metadata-based security approaches will ensure that data can be handled using the correct security approaches and engagement modules. The use of the keen approaches to define data management and handling will craft a mechanism that ensures an increasingly beneficial classification of data to address the needs as they come into the company. Nonetheless, tagging sensitive data will provide encryption, protection, and compliance to ensure the scope of control and handling for the data to achieve suitable handling at all levels. These advances will create better data handling approaches to achieve the proper outcomes in backup modeling and address.

d) **Homomorphic Encryption:** Modern companies seek to exploit the use of computational handling of homomorphic encryption to ensure that it can be maintained during the recovery and backup address process. The enhanced security of the data platforms creates a reliable way to ensure that their processing and analysis are critically handled to achieve the most remarkable way of addressing data deployment and handling in the organization.

e) **Confidential Computing:** The use of trusted execution environments (TEE) creates a reliable step to ensure that data is encrypted during the processing phase. This ensures that isolated backup processes can run and secure the environments, protecting them from external threats at any given point in the handling and management of the information.

f) **Granular Backup and Recovery Services (BaaS):** This approach works based on capacity to ensure critical controls to help achieve proper backup handling. The use of service-level agreements (SLAs) will ensure compliance and lead to better handling of self-service portals, which will encourage and provide appropriate modeling of the data categories to achieve the best remarks in backup proficiency. Thus, these developments ensure that data can be addressed to achieve the most efficient level of backup provision.

g) **Unified Data Protection Platforms:** Using unified platforms will ensure critical steps are followed to manage and handle backup, disaster recovery, and management approaches to achieve the right outcome. These platforms will ensure essential management of the data across various platforms to achieve the highest result in any engagement delivered to them. Thus, the unified protection platforms will create a chance to ensure the data achieves the most remarkable deployment to address the organizations' core needs at any given point.

Summary

This chapter discusses backup and restoration approaches to handling ransomware attacks. The chapter begins with an instrumental analysis of backup types and the importance of backup in organizations. The different types of backups used in modern times include full, differential, and incremental backups. The chapter indicates situations when each backup model can be used, illuminating the importance of each, and the demerits that must be addressed for every backup approach. Additionally, this chapter states that the benefits of using data backup are the capacity to restore information after a ransomware attack, ensuring business continuity, enabling mitigation for ransoms, and achieving the proper scope of detailing engagement within the business at any point. Different backup practices can be applied to ensure the proficiency of backup management and modeling techniques. In the first instance, developing a way to define and work with approaches such as geographical locations, emergency recovery kits, and incremental strategy, the backup practices highlight beneficial steps that can appeal to managing data backup needs at any given level. Using data backups also calls for innovative integration of modern technologies such as data-centric security, unified platforms, and AI/ML integration in data backup management and handling systems.

References

[1] F. Faisal, "The backup recovery strategy selection to maintain the business continuity plan," *Journal of Applied Sciences and Advanced Technology*, vol. 1, no. 1, pp. 23-30, 2018.

[2] D. Min et al., "Amoeba: An autonomous backup and recovery SSD for ransomware attack defense," *IEEE Computer Architecture Letters*, vol. 17, no. 2, pp. 245-248, 2018.

[3] K. Lee, S. Y. Lee, and K. Yim, "Machine learning based file entropy analysis for ransomware detection in backup systems," *IEEE Access*, vol. 7, pp. 110205-110215, 2019.

[4] J. Thomas and G. Galligher, "Improving backup system evaluations in information security risk assessments to combat ransomware," *Computer and Information Science*, vol. 11, no. 1, 2018.

[5] G. Ramesh, J. Logeshwaran, and V. Aravindarajan, "A secured database monitoring method to improve data backup and recovery operations in cloud computing," *BOHR International Journal of Computer Science*, vol. 2, no. 1, pp. 1-7, 2022.

[6] D. Chang, L. Li, Y. Chang, and Z. Qiao, "Cloud computing storage backup and recovery strategy based on secure IoT and spark," *Mobile Information Systems*, vol. 2021, pp. 1-13, 2021.

[7] H. Rezaeighaleh and C. C. Zou, "New secure approach to backup cryptocurrency wallets," in *2019 IEEE Global Communications Conference (GLOBECOM)*, 2019, pp. 1-6.

[8] H. Adkins et al., *Building secure and reliable systems: best practices for designing, implementing, and maintaining systems*. O'Reilly Media, 2020.

[9] L. Corti, M. Woollard, L. Bishop, and V. Van den Eynden, *Managing and sharing research data: A guide to good practice*, 2019.

[10] G. G. Gueta et al., "SBFT: A scalable and decentralized trust infrastructure," in *2019 49th Annual IEEE/IFIP International Conference on Dependable Systems and Networks (DSN)*, 2019, pp. 568-580.

[11] E. Eryurek, U. Gilad, V. Lakshmanan, A. Kibunguchy-Grant, and J. Ashdown, *Data governance: The definitive guide*. O'Reilly Media, Inc., 2021.

[12] R. Anderson, *Security engineering: a guide to building dependable distributed systems*. John Wiley & Sons, 2020.

[13] F. M. Awaysheh et al., "Security by design for big data frameworks over cloud computing," *IEEE Transactions on Engineering Management*, vol. 69, no. 6, pp. 3676-3693, 2021.

[14] B. Seth et al., "Integrating encryption techniques for secure data storage in the cloud," *Transactions on Emerging Telecommunications Technologies*, vol. 33, no. 4, p. e4108, 2022.

[15] V. Shah and S. R. Konda, "Cloud computing in healthcare: opportunities, risks, and compliance," *Revista Española de Documentación Científica*, vol. 16, no. 3, pp. 50–71, 2022.

CHAPTER 12

Ransomware Recovery Framework

Author:
Anirudh Khanna

Introduction to Ransomware Recovery

Ransomware is a cyber threat that targets a company's computer system and demands a ransom for releasing data or unlocking accounts. Ransomware attacks are currently considered among the most significant threats to companies of any size, type of organization, or field of activity. These stealthy malware threats can easily secure essential contents and applications, bringing a company's functionality to a halt until the price for their release is met or, in other words, decrypted. Developing a robust and comprehensive recovery plan for ransomware is critical because it will allow one to limit the amount of time spent and data lost in the process. This chapter gives detailed information about the step-by-step procedure for creating and enforcing an efficient and comprehensive recovery plan for ransomware attacks. It describes an entire process, from identifying an attack to a system rebuild and configuration adjustments that discourage the attackers. Further, it discusses methods and techniques for quickly identifying and preventing infected computers from leaking encryption across the network.

© Saurav Bhattacharya 2024
A. Khanna, *Securing an Enterprise*, https://doi.org/10.1007/979-8-8688-1029-9_12

CHAPTER 12   RANSOMWARE RECOVERY FRAMEWORK

As with any cyber-security event, each organization needs to have an outlined plan of action to be taken when combating ransomware attacks. Rapid response and strict compliance with these measures can help avoid a brief interruption and a severe data loss catastrophe [1]. Figure 12-1 gives an overview of the process.

```
[Start] --> [Identify the attack] --> [Isolate Infected Systems] --> [Assess the
Damage] --> [Report the Incident] --> [Identify the Ransomware Variant] --> [Decide
on a Course of Action]
|
|----> [Preserve Evidence]
[Decide on a Course of Action] --> [Restore from Backups] --> [Decrypt Files (If
Possible)] --> [Rebuild and Harden Systems] --> [Review and Improve Security] -->
[End]
```

Figure 12-1. Ransomware Recovery Process Flowchart

It is crucial to note that, as shown in Figure 12-1 and explained in this section, whether the "Restore from Backups" and "Decrypt Files" procedures are performed depends on the decision made regarding the best course of action. Should the choice not be made to decrypt or restore, the method will proceed straight to "Rebuild and Harden Systems."

Identify the Attack

The first significant action that should be taken is to assert with certainty whether the system has been infected with ransomware. Public awareness is the key since data encryption can quickly go out of hand, as evidenced by the events experienced in the recent past. Common indicators of a ransomware attack include:

- Files in the network are afflicted with encryption, which is frequently followed by unfamiliar file suffixes.

CHAPTER 12 RANSOMWARE RECOVERY FRAMEWORK

- A ransom message is displayed on one or two of the system screens, including the files' decryption and the required amount to be paid [2].

- It begins with privileges to files, shared drives, databases, and any other resource being limited or access to them being denied.

- Tools for endpoint protection identify questionable encryption procedures and malicious activity.

- Unusual increases in network traffic happen as the ransomware spreads laterally.

In cases where the presence of ransomware is evident through clear indications, the organization's leaders have to declare a response level and activate the pre-existing plan. Ransomware has the potential to change course and inflict significant harm if attack verification and reaction are delayed [3].

Isolate Infected Systems

When a ransomware attack is established, some actions have to be taken to limit the spread of the malware to other systems and sever its connectivity to the production network [4]. Unlike the representations of the threat lying horizontally encrypting data across a larger area, this containment measure stops its spread in that direction. Specific isolation actions include:

- Physical removal of Ethernet cables or setting infected nodes into a state where they no longer have active Wi-Fi radios [4].

- Porting or filtering out the particular IP address at the Firewall or router level when the address has already been found to be tapped by hackers.

CHAPTER 12   RANSOMWARE RECOVERY FRAMEWORK

- Temporarily suspending all the services running on the internet and affected computers.

- Disconnecting all unnecessary hosts and services, cloud services, and processes operating from contaminated computers

These steps suspend and stop any further encryption by this ransomware or data leakage to other areas of the network. Thus, when the foothold is contained, the response teams can assess the infection systematically.

Assess the Damage

After that, cyber respondents should thoroughly evaluate the specific impact of encryption as part of the ransomware attack. Key damage assessment activities entail:

Determining all the systems or sources of data subjected to encryption

- Estimating the amount of data that is encrypted based on file and storage space utilized

- Ascertaining if any cloud services, backups, or off-site data was also encrypted [5].

- Identifying the type of ransomware by analyzing its features, including the ransom notes.

- Learning whether there was any unauthorized transfer of data by the attackers before encryption

This enables the identification of the potential exposure to data loss and, thus, its impact on the organization's operation, which can inform further action during the damage assessment phase.

CHAPTER 12   RANSOMWARE RECOVERY FRAMEWORK

Report the Incident

Following the initial isolation and assessment, the cyber security incident should be recorded, and proper notification to other internal stakeholders should be given as provided by the organization's cyber incident response procedure. Report details should cover the following:

- Recent attack indicators and time frames, starting from the first time the attack was noticed to the present time the attack was last dismantled [5].
- A list of the affected systems, programs, and data storage places that are currently known.
- Common name/variant of the strain, if known.
- Calculated based on damage assessments of encryption amounts.
- Details of the Ransom payment demand, if there is any.
- Measures that have been taken so far not to aggravate the problem and loss of data.

This internal report enables the key leaders and decision-makers to be updated depending on the progress made in the recovery process. For many organizations, ransomware incidents also need to be reported externally to

- Law enforcement agencies, which are responsible for criminal investigation in the event of an incident [6]
- Related trade associations, as necessary
- Cyber insurance providers, if a policy has been taken
- Customers and business partners who are affected as per the legal data breach notification laws

CHAPTER 12 RANSOMWARE RECOVERY FRAMEWORK

Identify the Ransomware Variant

Based on the recommendations to cyber responders included in the incident analysis, it is crucial to determine the specific ransomware type used in the attack with certainty [7]. Characteristics like:

- Family name, if known, to which the specific ransomware belongs, for example, WannaCry, Locky, Ryuk, etc. [5].

- Files that are targeted for encryption based on the file extension.

- A ransom note usually contains content and wording focusing on payment and instructions on how to make the payment.

- Any other pertinent character attributes, like the PC's background, the different tune set for e-mail received, etc.

Finding the variation is essential since it determines whether remedial techniques or malware-specific decryption solutions can be applied. If the conflict results in an inconsistency, then organizations must choose a course of action.

Decide on a Course of Action

In this part, organization leaders are presented with the challenging task of making proper decisions on the most appropriate action plan to take to mitigate the threat and recover data where necessary. Options may include:

234

CHAPTER 12   RANSOMWARE RECOVERY FRAMEWORK

1. *Pay the Ransom (Not recommended)*

   Giving in to the attackers' monetary demands for data decryption, using a decryption tool or keys. This can be deemed the last resort when backups are unavailable, and the data is highly vital [8].

2. *Attempt Data Recovery*

   If backups are available, work can be done on the backup and recovery of the evidence from the known good copies. If the backups do not exist, then cybersecurity firms can obtain the data by other means.

3. *Rebuild Systems*

   Reimage and completely erase affected systems using reliable sources to remove any residual malware persistence methods [9]. Restore data from backups to these fresh systems. The following decision should consider issues such as budget, the importance of encrypted data, legal/compliance consequences of data leakage, the cost of inaccessibility, and decryption feasibility.

4. *Preserve Evidence*

   However, if the decision to reject the ransom payment is reached, then maintaining the integrity of the attack becomes relevant for future legal actions or investigations of the incident. Evidence sources may include:

   - Disk pictures and memory grabs from compromised endpoints
   - Traffic logs and captures from networks

- Malware samples and other host/network artifacts
- The text files and other related files with a ransom note
- Impacted user/system documentation

If such evidence is to be used in the future, it is recommended that it be appropriately preserved and handled according to the provisions of the chain of custody.

Restore from Backups

In organizations with well-developed data backup solutions, the next significant task is to recover data from these protected sources for newly reconstructed systems [10]. Key backup restoration activities include:

- Ensure the backups' authenticity and confirm that they were not encrypted.
- Sourcing from known clean sources if backup systems and storage need to be restored.
- Testing first in abstract configurations and then in general production recovery.
- Restore the backups in a structured manner according to the schedule formulated in the recovery plan.
- Verifying the restoration of data successfully using checksum and sampling

Restore points are the chief source of restoring a system after ransomware damage without yielding to the asked ransom. Regarding current backups, if they do not exist or are not feasible, potential ransomware decryption might be under consideration as another possible way to restore data. This involves performing decryption activities using third parties, resulting in attackers not being paid. From time to time, decrypted samples or specific decryption keys or utilities get released by law enforcement, cyber-intelligence companies, and security researchers [11]. These may free the encrypted data in case ransomware implementation has shortcomings or vectors that were leveraged. However, decryption is only sometimes possible or guaranteed, as already indicated above. The stronger and more robust the encryption implementation, the more work that would need to be decrypted to get the victims' materials.

Rebuild and Harden Systems

Once the data has been recovered through backups, decryption, or other means, the final step of the recovery process is to rebuild clean systems and implement the hardening controls [12]. This encompasses:

- Use reliable, current operating system (OS) images to wipe and re-image affected systems.
- Using trusted sources to download the latest applications and patches.
- Updating all the known patches/updates of operating systems, their Applications, and firmware.
- Applying specified security settings from different standards, such as CIS.
- Implementing fully updated solutions such as endpoint protection, DNS filtering, and other security solutions.

This is a good chance to remove the malware persistence from the environment and recode the vulnerabilities that allowed the first compromise.

CHAPTER 12   RANSOMWARE RECOVERY FRAMEWORK

Review and Improve Security

The final step is the analysis of organizational aspects, weaknesses, and opportunities that the ransomware infection can utilize after recovery. Some of these are the entrance point of the virus, any control gaps exploited, whether sufficient security measures and user awareness have been employed, and signs that point toward possible inadequacy in backup plans, DRP, cyber insurance, tools, and staff [13]. Outcomes should include a solid quality improvement plan that will help address the risks and enhance protective measures against new threats of ransomware attacks. Based on the conclusion drawn from the post-incident ransomware, an improvement plan to improve the institution's cybersecurity should be developed. Lessons learned to inform critical areas to prioritize vulnerability management by designing and implementing strict patch management procedures that require updating all the firm's systems, applications, and firmware whenever a new security patch is released. Unpatched application vulnerabilities serve to deliver ransomware into the systems. Another element that also needs to be addressed is managing the principle of least privilege through precise access controls due to how ransomware exploits existing permissions [14]. Segmentation of the network into specific security areas and strict traffic rules might help prevent ransomware spread. Organizations should put into practice EDR solutions that include behavioral analysis, anti-malware, data encryption, and other host functions. Security awareness training is critical, encompassing training programs for personnel on phishing threats, safe browsing, and social engineering techniques used by attackers in ransomware attacks [15].

Further, the backup systems need the same assessments and enhancements regarding redundancy, data availability, encryption, and air-gapping for ransomware recoverability. Moreover, cyber incident response plans must be enriched by detailed ransomware response procedures for the organization across all its departments. Therefore, one must employ a multi-layered improvement strategy to avoid future ransomware risks.

CHAPTER 12   RANSOMWARE RECOVERY FRAMEWORK

Identifying and Isolating Infected Systems

Quick identification of systems infected by ransomware and their separation from the rest of the network is essential for a successful mitigation process. The faster the spread of malware is stopped, the fewer files may be at risk of being encrypted.

Network Segmentation

Before the internet was even a thing, network segmentation techniques started to take shape as firewalls and zone-based security techniques were created to shield networks from outside threats and illegal access. Although segmentation has become more complex and sophisticated due to cloud computing and virtual networks, many of the same fundamental ideas continue to stand true;

Zero-trust approaches: A zero-trust security approach is predicated on the idea that no one should be trusted, not even computers, when they ask to access a network or any of its applications; instead, they must authenticate themselves and justify their request. Zero-trust strategies identify network assets that need extra protection, laying the groundwork for segmentation [18].

> *Divided networks:* After the necessary data and resources have been identified and the optimal borders have been created, several distinct network segments can be made. Two popular methods for network subdivision include using subnets to split the network based on IP addresses or utilizing virtual local area networks (VLANs) to construct a collection of smaller, virtually connected networks.

CHAPTER 12   RANSOMWARE RECOVERY FRAMEWORK

Microperimeters: When network segmentation security is implemented, conventional concepts of what is within and outside the network are altered. Segmentation-created micro perimeters provide distinct perimeters for each app or service. This more detailed level of customization is available for security measures like authentication [18].

Establish access controls: After the network has been divided into sections, access controls need to be created for each section to ensure that the correct authentication procedure is followed when granting access and making traffic flow policies. The least privilege principle expands on zero-trust tactics by granting access to programs or gadgets only when necessary for an individual to carry out their duties.

Endpoint Detection and Response (EDR)

EDR solutions like Microsoft Defender, CrowdStrike, and SentinelOne consolidate multiple endpoint capabilities, including

- Enduring behavior and consistent watching and analyzing
- Other strategies for attack identification including machine learning for malicious encryption, process injection, and the like
- Automated threat hunting, as well as response actions such as file or process killing
- Swamp analysis by centralized data collection

CHAPTER 12 RANSOMWARE RECOVERY FRAMEWORK

EDR is a solid first layer of protection against ransomware delivery and host activity.

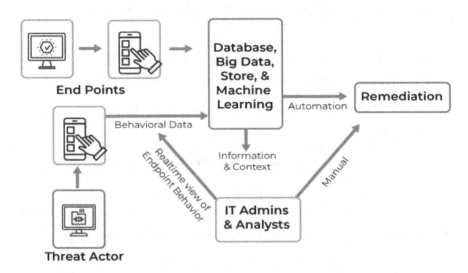

Figure 12-2. *The Mechanism behind EDR Performance [17]*

From Figure 12-2. Endpoint Detection and Response (EDR) systems are defined as Information Technology solutions that detect threats within endpoint devices. EDR agents are deployed on endpoints and gather information about their activity, which is transmitted to the EDR management server that processes the data with AI and ML [17]. EDR employs different threat detection mechanisms such as signature analysis, behavioral analysis, analysis of suspicious activities in a sandbox, and other tests such as comparison of activities against a white- or blocklist [17]. When a threat is identified, the EDR sends an alert, and in some instances, the EDR system retaliates against the danger. This allows stakeholders to follow through with relevant actions such as quarantining or deleting inconvenient and dubious files, quarantining endpoint devices, and exploring the situation.

CHAPTER 12   RANSOMWARE RECOVERY FRAMEWORK

Intrusion Detection/Prevention (IDPS)

Computer network and host-based IDS/IPS systems detect and prevent a system's traffic and activities containing identified malicious patterns. Signatures can be configured to alert on ransomware network behaviors like

- Abnormal levels of Server Message Block or SMB connections that are associated with lateral movements
- Increase in DNS requests to the domains of interest linked to command-and-control
- TOR traffic or usage of anonymization services by internal assets

Depending on the kind of deployment they want, organizations might consider implementing any one of four types of IDPS.

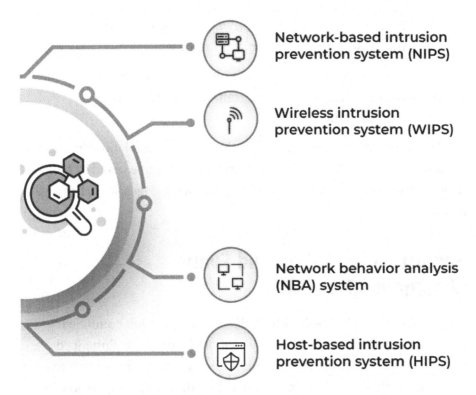

Figure 12-3. *Types of IDPS [16]*

As Figure 12-3 presents, there are several types of Intrusion Detection and Prevention Systems (IDPS), each with its own focus and deployment location: As Figure 12-2 presents, there are several types of Intrusion Detection and Prevention Systems (IDPS), each with its own focus and deployment location:

> *Network-based intrusion prevention system (NIPS)*: Scans the whole network for malicious traffic and analyzes protocol activity.
>
> *Wireless intrusion prevention system (WIPS)*: Supervises wireless networks by analyzing wireless networking protocols.

CHAPTER 12 RANSOMWARE RECOVERY FRAMEWORK

Network behavior analysis (NBA) system: Checks for risk by scanning for traffic abnormalities.

Host-based intrusion prevention system (HIPS): Used with one host to supervise traffic and activities on that host [16].

The selection of IDPS depends on the organizational infrastructure, the required size of the organization, and the techniques used by IDPS to detect intrusions. Regarding the active encryption stages, IDPS has a part in identifying precursor activities.

Security Information and Event Management (SIEM)

SIEM solutions such as Splunk, QRadar, and LogRhythm combine log data from networks, endpoints, cloud services, and more. Enhanced and selectively constructed correlation rules, reports, and analytic use cases embedded in this centralized data identified as having to do with ransomware indicators can fast-track the triaging process of potential threats. SIEMs serve as the central analyzer, working with tools like:

- EDR
- Web proxy logs
- Cloud access security brokers (CASB)
- IDPS
- Data loss prevention (DLP)

Table 12-1. *Differences between various SIEM solutions: Splunk, QRadar, and LogRhythm*

Feature	Splunk	QRadar	LogRhythm
Data Collection	Collects data from virtually any source (networks, endpoints, applications, etc.)	Collects data from networks, endpoints, logs, flows, vulnerabilities, etc.	Collects data from networks, endpoints, cloud services, applications, etc.
Data Analysis	Uses Search Processing Language (SPL) for data correlation and analysis	Uses QRadar Correlation Engine for data analysis and correlation	Uses AI Engine for data analysis, correlation, and threat detection
Threat Detection	Uses machine learning and behavior analytics for threat detection	Uses rules, risk-scoring models, and behavior analytics for threat detection	Uses User and Entity Behavior Analytics (UEBA) and machine analytics for threat detection
Incident Response	Provides incident review and investigation capabilities	Offers incident response and case management capabilities	Includes incident response and case management workflows
Reporting & Dashboards	Customizable dashboards and reporting capabilities	Pre-built and customizable reports and dashboards	Out-of-the-box and custom dashboards and reporting
Cloud Support	Available as a cloud service (Splunk Cloud) or on-premises	Available as a cloud service (QRadar on Cloud) or on-premises	Available as a cloud service (LogRhythm Cloud) or on-premises
Pricing Model	Primarily based on data ingestion volume	Primarily based on events per second (EPS).	Primarily based on data processing volume

User Behavior Analytics (UBA)

More advanced UBA tools like Securonix or Exabeam can also detect lousy user behavior or suspicious credential stuffing by learning the tenant's routine processes and user activity patterns [19]. Essentially, breaches that deviate too much from these baselines point toward credential theft, privilege escalation, and other early steps in a ransomware attack under compromised user contexts. Combining all these measures and using them as layers of improvement is a way to foster the strengthening of the organization's defenses against future ransomware threats. Besides, the implications of the post-incident ransomware review should be used in formulating an overall improvement in a cybersecurity strategy. The following are the highlighted areas where organizational learning from past breaches can be applied for improved risk management: This can alert hackers to unpatched vulnerabilities through which the ransomware can be delivered. Another aspect is enforcing the principle of least privilege with the help of narrow access rights since ransomware thrives in environments that allow it to address excessive permissions [20]. Topologically, dividing the network into several security areas with limited traffic can curb ransomware propagation. Organizations should use various endpoint detection and response (EDR) solutions incorporating behavioral analysis, reactive malware, data protection, and other host functions.

Frequent security awareness training that informs people about distinguishing phishing, safe browsing, and social engineering strategies involved in ransomware attacks is priceless. In addition, practical inspection and enhancement of the backup systems in aspects of redundancy, data consistency, encryption, and air-gapping will make them capable of handling ransomware attacks. Also, CIDR programs need to be further defined with detailed ransomware response procedures covering processes throughout the organization.

CHAPTER 12   RANSOMWARE RECOVERY FRAMEWORK

Honeypots

Honeypots are the actual systems whose primary purpose is to attract the attackers' attention and simultaneously secure production targets [21]. For ransomware, these honeypots can be staged as

- Have deceptive file shares or databases that include the monitored honey files.
- Voluntarily exposed assets can initialize numerous interactions.
- Separated zones that, in their general outline, resemble production hosting.

As long as defenders can prevent anyone but the "attacker" from encrypting these lure files or accessing these isolated honeypot segments, it provides a safe space for observing the attackers' procedures, as depicted in Figure 12-4.

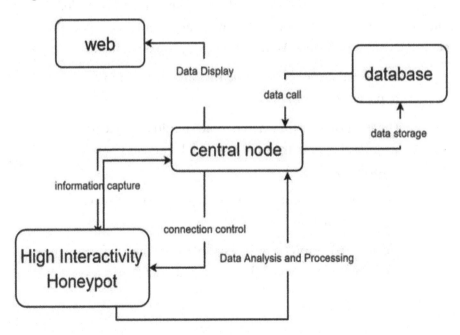

Figure 12-4. A representation of Honeypot [22]

In Figure 12-4, the parts of a high-interactivity honeypot system are designed to interact with and engage the attacker who is infesting the system. The honeypot itself is bait, whereas the true nature of the trap is the system, which is meant to resemble an authentic operating environment that will draw the attention of attackers. As the attackers interact with it, the honeypot has a hub function where all the connections are processed, and all the actions and commands of the attacker are observed. The obtained data is then transferred to database storage and can be subsequently analyzed for detailed information. It becomes relevant to examine the attacker's practices—to determine whether there are any weaknesses in the defender's infrastructure and to detect the malware employed, if any [22].

Incident Response Playbooks

Ensuring detailed playbooks exist for the response when a ransomware attack has occurred, as well as when the ransom demand is made, indicating the actions that should be undertaken and the responsible personnel to conduct them. Well-defined playbooks spell out:

- Tasks and duties concerning the different response actions. [22]

- Clear guidelines on when people should escalate to the next level of management or to the leader and how to inform them.

- Tools and data sources are required to be used as per listed scenarios.

- Automated enforcement measures are to be implemented at every phase (such as isolation stages).

CHAPTER 12 RANSOMWARE RECOVERY FRAMEWORK

Incident response playbooks serve as a united strategy within security operations, IT, crisis management teams, and top management. By introducing controls for monitoring an attacker's behaviors, rationalizing and utilizing deception techniques analytics, and establishing reaction protocols, an organization can obtain the needed time to immediately compartmentalize a ransomware incident before full encryption takes place [22]. Using these capabilities and analyzing a broad-spectrum recovery framework discussed in the section "Introduction to Ransomware Recovery" lays down a secure, multilayered strategy for mitigating ransomware risks.

Summary

Chapter 12 is a step-by-step guide that can help any organization prepare to respond to ransomware attacks with fewer negative consequences to its data and operations. It presents the multiple phases of the recovery process, including the identification of the attack, the containment of the system, the investigation into the incident, the acquisition of evidence, the data recovery, and the system rebuild.

Converting as soon as possible to the defined response procedures is mandatory in the early stages to prevent ransomware from spreading in the environment. As the chapter recommends, creating and updating the backups that can be used as the main path of turning to ransom without paying it themselves is crucial. As with backups and insurance policies, the chapter identifies possible decryption mechanisms and decision-making criteria for choosing the last resort option: ransom payment. It also focuses on the insight that organizations should strengthen prevention controls and minimize attack vectors after a breach.

CHAPTER 12   RANSOMWARE RECOVERY FRAMEWORK

The second part of the chapter is dedicated to the strategies and tools that can help to quickly locate and contain the infected nodes before the encryption algorithm reaches out to more systems. It includes the capability of network monitoring, the function of endpoint detection and response, behavioral analysis of users/entities, deception tools like honeypots, and the function of security information and event management solutions.

Ultimately, this chapter presents an authoritative resource providing reliable and actionable steps to minimize the serious organizational consequences of ransomware attacks and ensure organizations of various sizes can recover from these incidents. It enhances the generally specified concepts of prevention, supported through proper planning and the employment of varied layers of recognition and mitigation.

References

[1] K. Yasar, "What is Disaster Recovery (DR)?," *SearchDisasterRecovery*, 2023. https://www.techtarget.com/searchdisasterrecovery/definition/disaster-recovery

[2] R. S. I. Security, "How to Identify Signs of Ransomware Attacks," *RSI Security*, Oct. 11, 2021. https://blog.rsisecurity.com/how-to-identify-signs-of-ransomware-attacks/

[3] C. Beaman, A. Barkworth, T. D. Akande, S. Hakak, and M. K. Khan, "Ransomware: Recent advances, analysis, challenges and future research directions," *Computers & Security*, vol. 111, no. 1, Dec. 2021, doi: https://doi.org/10.1016/j.cose.2021.102490.

[4] National Cyber Security Centre, "Mitigating malware and ransomware attacks," *www.ncsc.gov.uk*. https://www.ncsc.gov.uk/guidance/mitigating-malware-and-ransomware-attacks#:~:text=post%2Dexercise%20activity.- (accessed Jun. 11, 2024).

[5] Perception-point, "How Ransomware Attacks Work: Impact, Examples, and Response," *Perception Point*. https://perception-point.io/guides/ransomware/how-to-prevent-ransomware-attacks/#:~:text=Assess%20the%20damage%3A%20Determine%20the (accessed Jun. 11, 2024).

[6] P. Cichonski, T. Millar, T. Grance, and K. Scarfone, "Computer security incident handling guide," *Computer Security Incident Handling Guide*, vol. 2, no. 2, Aug. 2012, doi: https://doi.org/10.6028/nist.sp.800-61r2.

[7] Flashpoint, "The Seven Phases of a Ransomware Attack: A Step-by-Step Breakdown of the Attack Lifecycle," *Flashpoint*, Jul. 10, 2023. https://flashpoint.io/blog/the-anatomy-of-a-ransomware-attack/

[8] iStorage, "Ransomware: Avoid paying a King's ransom for your data," 2022. Accessed: Jun. 11, 2024. [Online]. Available: https://istorage-uk.com/wp-content/uploads/2022/11/iStorage_Ransomware_Avoid-paying-a-Kings-ransom-for-your-data.pdf

CHAPTER 12   RANSOMWARE RECOVERY FRAMEWORK

[9]   Z. Gittins and M. Soltys, "Malware Persistence Mechanisms," *Procedia Computer Science*, vol. 176, pp. 88–97, 2020, doi: https://doi.org/10.1016/j.procs.2020.08.010.

[10] S. Suguna and A. Suhasini, "Overview of data backup and disaster recovery in the cloud," *IEEE Xplore*, Feb. 01, 2014. https://ieeexplore.ieee.org/abstract/document/7033804

[11] J. Lewis, D. Zheng, and W. Carter, "The Effect of Encryption on Lawful Access to Communications and Data F E B R U A R Y 2 0 1 7," 2017. Available: https://home-affairs.ec.europa.eu/document/download/8387f390-4fd7-4bca-9b56-c1a186b301c4_en?filename=csis_study_en.pdf

[12] B. Plankers, "Ransomware Defense and Recovery Strategies | VMware," *The Cloud Platform Tech Zone*, 2024. https://core.vmware.com/ransomware-defense-and-recovery-strategies (accessed Jun. 11, 2024).

[13] National Center for Education Statistics, "Chapter 5-Protecting Your System: Physical Security, from Safeguarding Your Technology, NCES Publication 98-297 (National Center for Education Statistics)," *Ed.gov*, 2019. https://nces.ed.gov/pubs98/safetech/chapter5.asp

[14] M. McCarthy, "Principle of Least Privilege (PoLP) | What is it & Why is it important | strongDM," *www.strongdm.com*, Sep. 13, 2023. https://www.strongdm.com/blog/principle-of-least-privilege

[15] B. Gardner, "Chapter 1: What Is a Security Awareness Program?," *ScienceDirect*, Jan. 01, 2014. https://www.sciencedirect.com/science/article/abs/pii/B9780124199675000016

[16] R. Mohanakrishnan, "What Is Intrusion Detection and Prevention System? Definition, Examples, Techniques, and Best Practices," *Spiceworks*, Feb. 11, 2022. https://www.spiceworks.com/it-security/vulnerability-management/articles/what-is-idps/

[17] C. BasuMallick, "What Is Endpoint Detection and Response? Definition, Importance, Key Components, and Best Practices," *Spiceworks*, Sep. 08, 2022. https://www.spiceworks.com/it-security/endpoint-security/articles/what-is-edr/

[18] Dashlane, "What is Network Segmentation & How Does It Work?," *Dashlane*, Feb. 22, 2024. https://www.dashlane.com/blog/what-is-network-segmentation

[19] Exabeam, "What Is UEBA (User and Entity Behavior Analytics)?," *Exabeam*, 2023. https://www.exabeam.com/explainers/ueba/what-ueba-stands-for-and-a-5-minute-ueba-primer/

[20] Cisco, "Cisco Application Policy Infrastructure Controller (APIC)—Cisco ACI Hardening," *Cisco*, 2023. https://www.cisco.com/c/en/us/td/docs/dcn/whitepapers/cisco-aci-hardening.html (accessed Jun. 11, 2024).

[21] SentinelOne, "What is Honeypot? | Working, Types, Benefits, and More," *SentinelOne*, 2024. https://www.sentinelone.com/cybersecurity-101/honeypot-cyber-security/#:~:text=Honeypots%20aim%20to%20mimic%20a (accessed Jun. 11, 2024).

[22] Atlassian, "How to create an incident response playbook," *Atlassian*, 2024. https://www.atlassian.com/incident-management/incident-response/how-to-create-an-incident-response-playbook

CHAPTER 13

Negotiating with Attackers

Author:
Anirudh Khanna

Overview

The issue of whether or not to engage in negotiations with ransomware attackers is not easy. It has many pros and cons regarding legal, ethical, and practical implications. Unfortunately, there is no straightforward answer to this question because the application of these measures may depend on the existence and characteristics of the attack, as well as on the organization that is being targeted. This chapter will, therefore, explain the underlying factors regarding negotiations and inform organizations of some of the best practices for dealing with cyber criminals. Ransomware is a constantly evolving threat that targets businesses, governments, and individuals in the form of malicious software that demands that the victim pay the attacker a sum of money within a specified time for the decryption key [1]. The mobile malware encrypts data retrieved from a vulnerable device and then holds this data for ransom. Disaster recovery is critical to virtually any organization, given the continuous pressure placed on organizations today to restore their lost data and systems as much as possible and in the shortest time possible to avoid significant losses. It may be tempting to deal with cyber criminals to 'get the job done' quickly, but it poses questions that are pretty difficult to answer [2]. In other words, it could be seen that paying for the ransom might be equivalent to financing the criminals, and this might contravene legal prohibitions or restrictive measures. There are arguments concerning using funds that hackers and similar undesirable elements would use to finance such malice toward others. Paying the ransom may lead to even more reckless attacks since the perpetrators know this is an effective way to get what they want. However, as they decide not to pay the ransom, most of the victims may lose data permanently, which may lead to the crippling of operations for good. This chapter investigates these compounding factors to provide leadership direction for the process. In considering potential and actual issues and dilemmas, we consider real-life cases, legal provisions and the common law, ethical theories, and opinions from practitioners and scholars. Thus, the above-mentioned complex analysis aims to provide the toolkit that would help organizations analyze their particular case and make the proper decision rationally, as well as minimize the motivations for future ransomware activities.

Key Statistics on Ransomware Attacks in the United States, Europe, and Globally

The threat of ransomware attacks has increased rapidly globally in the past few years. In the USA, ransomware reports increased by 60% to 2,474 in 2020, as recorded by the Internet Crime Complaint Center belonging to the Federal Bureau of Investigation. This survey revealed the reported ransomware attack cases, with the totality of the victim losses amounting to more than $4.2 billion, as presented in Figure 13-1 [3]. However, this may not even be 10% of the bill, as several victims do not report the circumstances to law enforcement.

CHAPTER 13   NEGOTIATING WITH ATTACKERS

Figure 13-1. *FBI Report on Ransomware Attack in 2020 [3]*

Even organizations researching and monitoring ransomware attacks and their impact have presented a worse picture. Cybersecurity Ventures predicted that global ransomware was worth 20 billion U.S. dollars in 2021 and would rise to over 42 billion periodically over the years. They anticipate that ransomware costs can increase to $265 billion per year by the year 2031 if it is not checked and to over 9.5 trillion globally in 2024 [4]. Ransomware is a severe public sector problem in the United States, affecting state and local governments, schools, hospitals, and infrastructure operators. Local governments, universities, and schools were hit by ransomware in 2021 at a rate of over 230, as reported by the cybersecurity firm Emsisoft. Some of the most high-profile cyberattacks entailed the Colonial Pipeline and Ireland's healthcare sector, which took months to resolve [37].

Europe has not been any different, with high-profile ransomware attacks affecting the continent. More than 60% of ransomware attacks on organizations in 2021 were based in Europe, according to the EU Agency for Cybersecurity [5]. Several nations, including the UK, France, Germany, and Italy, were among the most affected. The WannaCry case was one of

257

CHAPTER 13   NEGOTIATING WITH ATTACKERS

the most devastating global ransomware attacks in 2017, with over 200,000 systems affected in over 150 countries [6]. For this reason, the highest economic cost is estimated to have been incurred by the UK's National Health Service, whose overall consumer costs exceed £92 million ($115 million) [7].

Looking more broadly worldwide, the anti-malware software provider Emsisoft analyzed global trends and found that.

a. The combined public and private expenses of all ransomware attacks in 2021 reached $20 billion.

b. The number of ransomware operators grew, with more than a hundred notable strains active.

c. The average demanded ransom increased by 45% to $75 billion by 2021 [8].

d. Lost time due to ransomware attacks represented 24 percent of the total expense.

Table 13-1. Ransom Demand Costs Breakdown in Various Countries. By 2021, the average ransom demand had risen to $75 billion [8]

Country	Total Submissions	Minimum Cost (USD)	Estimated Costs (USD)
United States	23,661	$920,353,010	$3,682,228,067
Italy	9,226	$346,729,130	$1,387,389,097
Spain	8,475	$298,254,459	$1,193,709,500
France	7,824	$283,816,080	$1,135,795,109
Germany	7,138	$252,609,210	$1,011,001,498
UK	4,788	$169,182,845	$677,113,461
Canada	4,257	$164,772,274	$659,246,267
Australia	2,775	$105,978,531	$424,034,780
Austria	1,254	$46,643,868	$186,645,857
New Zealand	399	$14,230,333	$56,951,495
TOTAL (All countries)	506,185	$18,658,009,233	$74,632,036,933

The attackers have not limited themselves to large organizations and enterprises; IT companies and consumers are now in the loop. A cybersecurity firm, Heimdal Security, estimates that global ransomware attack attempts stood at 304.7 million in the first six months of 2021 alone [9]. Such statistics indicate that ransomware has evolved from a localized menace to a pandemic that cuts across industries and regions. Eradicating ransomware will require a concerted effort from governments and enterprises alike to employ dissuasion, mitigation, and dismantling of the monetary incentives that skew these cybercriminals' undertakings.

CHAPTER 13   NEGOTIATING WITH ATTACKERS

Legal and Ethical Considerations

Engaging in a negotiation process with ransomware attackers may raise several legal and ethical questions. On the other hand, the decision to pay the ransom looks rather tempting, as it is necessary to recover the encrypted data and continue work. However, this would encourage unlawful activity and violate laws prohibiting the funding of such actions. Payment does not ensure that files will be decrypted or that data leaks will be avoided [10]. It is somewhat unethical to pay ransoms because this prolongs the existence of ransomware and may encourage even more attacks on other victims [13]. Organizations must consider possible risks and ethical issues and whether encouraging crime pays off, considering the need to regain data quickly. Due to ransomware attacks' highly technical and dangerous nature, negotiation's legal and moral analysis is complex.

Legal Considerations

The legal implications of paying cyber ransoms are critical, and therefore, organizations face these challenges with potential legal repercussions. Paying ransomware attackers avails resources and money to individuals involved in criminally related activities such as stealing data, dealing with cyber extortion, and other related vices. This entails a whole lot of legal risks that extend across different jurisdictions. In the USA, the Criminally Derived Property Statute prohibits anyone from using the realizable amount of money known to have been acquired through the commission of certain federal crimes, such as hacking, fraud, or the unlawful retention of data [11]. The ransom payment requests stem from actual cybercriminal activities. Therefore, any payment sent would amount to money laundering. The US government has recently directly informed the private sector that providing help with ransomware payments may lead to violations of the Anti-Money Laundering Laws [12, 13]. Further, the

CHAPTER 13   NEGOTIATING WITH ATTACKERS

U.S. Treasury Department's Office of Foreign Assets Control agency also has sanctions restricting dealings with cyber criminals and related persons [13]. Another strain of ransomware, such as Ryuk, SamSam, and Dridex, is associated with the sanctioned Russian cybercrime groups [14]. Direct or indirect payments to these sanctioned entities by paying for a ransomware attack expose the organization to enforcement actions that may lead to fines. Thus, the U.S. government made a giant step in reacting to the ransomware threat in 2021 when OFAC imposed the first ransomware-specific sanctions against the Russian-based cryptocurrency exchange SUEX OTC. This indicated that ransomware payments are within the regulators' sights as priorities move from essential response to active prevention. In 2020, specifically in January, the creators of the Dridex banking Trojan unleashed a highly damaging ransomware attack on hundreds of organizations. One of the affected firms was a US-based company that offered healthcare services. The firm had to succumb to the attackers' demands and settle for $2 million [15]. However, the FBI later stated that it had been funded as a branch of Evil Corp., a Russian cybercrime organization. This payment possibly evaded OFAC sanctions, thus putting the victim at risk of future enforcement [16]. As it goes with the regional laws in many other countries, any such funding of illegitimate activities is prohibited through Acts such as anti-money laundering and anti-terrorism financing laws. Globally, law enforcement has pointed out that paying ransomware leads other criminals to additional profitable revenue models for other crimes.

In 2016, a web hosting company in South Korea was attacked by ransomware, and they paid about 1 million dollars for decryption of the infected files' decryption keys. Still, the decryption keys were not functional [17]. Korean authorities later sanctioned the company for approximately $1 million by passing money laundering laws that ban the movement of cash that originates from criminal activities. In other words, paying the ransom and indirectly encouraging the work of the specific ransomware group, even when it is not under sanction, directly

261

contributes to the financing of these criminally derived funds, which can only lead to more destructive and crippling attacks [13]. There are no guarantees of successful data recovery or of ensuring that attackers will not leak stolen files after paying for them. From a specific liability perspective, organizations must consider whether ransomware payments could be associated with other unlawful activities, such as terrorism, human trafficking, or drug trafficking [18]. The victims could then be subjected to further legal proceedings, either locally or internationally, for funding the said crimes through the ransoms paid. Attribution of attacks is sometimes ambiguous, which we as organizations may fund as some of the most serious threats to national security. It is crucial to note that the legal risks of negotiating ransomware are complicated. Yet, the main aspect of mitigation is to evaluate the risks in detail and diffuse the prospects of any payment applicable under the law. Maintaining legal compliance with the police and other governing bodies helps avoid preconceptions regarding decision-making processes that reduce risks.

Moral Considerations

In the case of ransomware, the moral dilemma of negotiating with attackers to pay the ransom is not simple. On the other hand, there are corporate and prosaic needs to urgently regain the encrypted data and retaliate without a second thought. However, this has to be put against larger, more significant moral issues that make ransomware payments a moral dilemma. In some of the most basic concepts, some ransoms are paid to encourage and promote criminal activities that originate from the illegitimate invasion of computers and the theft of information [18]. Such a model generates new sources of income to attract more cybercriminals into the ransomware business. This sustains and aggravates an unethical model based on the generation of fear and the expectation of a payoff for not disclosing the stolen information. Each dollar paid out sustains that model's effectiveness. From an ethical point of view in terms of the

CHAPTER 13   NEGOTIATING WITH ATTACKERS

efficiency of the general interest, in simplified terms of Bentham, the ransom payments of ransomware generate negative externalities that decrease the welfare of society. Large payments to the ransom merchants by producing powerful, functioning, and well-established brands inform other criminals in the shady world that ransom businesses are lucrative to invest in. This increases the attractiveness of more skilled developers and infrastructure to improve future strains. This just fuels another cycle of rewarding the attackers with enhanced attacks that are even more sophisticated and destructive. He also acquires funds to popularize other related criminal businesses, including human smuggling, drug dealing, contract cybercriminals, and funding of terrorism. Thus, although one organization's payment may be justified on ethical grounds in a given organization, the consequences in the real world consistently promote further harm to society. Therefore, the ransomware gang's first action violates the Hippocratic Oath principle, "First, do no harm" [19].

For example, one of the most significant ransomware attacks began in 2021 by the group REvil, which infected hundreds of companies globally based on taking advantage of an existing flaw in the software of the targeted corporate entity [20]. One of the most revealing victims was JBS Foods, a Brazil-based company and one of the world's largest meat producers. JBS, the Brazilian meat-packing company, experienced a cyberattack targeting its data. To prevent further interruption to food supply chains, which the pandemic has shaken in the first place, the company's management agreed to pay $11 million to the attackers, one of the highest amounts of ransom disclosed in such a cyberattack [21]. While this payment was meant to reduce the operational disruption of the attack, it also funded other aspects of the operation and the maliciousness of REvil. Soon, REvil went a step ahead and began a more vicious rampage of attacks, facilitated by the $11 million funding from JBS [22].

It also raises more challenging issues, such as collaboration and the moral hazard of supporting certain regimes. An organization that can throw out the first attack or hacktivist while reaching out to pay the criminals retains ethical 'clean hands.' Nevertheless, if the victim did not apply reasonable measures that might have hindered the hackers or hacktivists from perpetuating their criminal act, it would be illogical not to negotiate with the criminals they have empowered. This is because it promotes poor security practices since users will act recklessly knowing that this main password protects their accounts. The ransomware payments also cause ethical dilemmas based on data importance instead of focusing on people's well-being [23]. A hospital, for example, is a typical example of a public sector victim who may be under pressure to pay ransoms to quickly regain the functionality of essential operations, which may threaten the lives and well-being of the public. However, in doing so, they take resources that otherwise could be used to support their primary objectives of promoting health, education, and community welfare. There are ethical issues concerning accessory liability and possibly even prosecution due to testing. Evaluating current legislation in many jurisdictions leads to the conclusion that it is unlawful for anyone to receive any benefits from criminal operations and to provide facilities for transferring funds of criminal origin. Thus, victims who bargain for payments may face ethical and legal compensatory liability risks [24]. Even though each situation must be considered individually, the moral ambiguities surrounding ransomware highlight why many governmental organizations advise against making payments wherever feasible. In the long run, the most effective method to starve and disincentivize the unethical ransomware ecosystem is to prioritize societally ethical behavior over short-term interests despite the overwhelming difficulty of doing so.

Balancing Risks and Benefits

Negotiating with ransomware attackers puts organizations in a tricky financial gain/loss analysis where there is no perfect right thing to do. Meeting the ransom demand is a positive scenario as it provides opportunities to quickly decrypt the systems and recover data to normalize functioning. This could help avoid other adverse impacts such as increased operation losses, rectification costs, damage to reputation, and other effects that may result from prolonged disruption [25]. For some organizations, faster data recovery could be a survival issue for the business to continue operations. Depending on time, reviving data from the backups or consistently promoting it costs more than the ransom and still does not result in a full recovery. The fact is that payments can be deemed to be the most realistic solution that can be implemented as long as many different ethical and legal risks have been identified.

On the other hand, accepting ransomware terms makes the extortion-based criminal business model legitimate, which means that one is funding cyberattack infrastructure that will later be used on other victims. It empowers the offenders and generates other social consequences that are more severe and permanent than the direct outcome. In many circumstances, even after the attackers have received a payment, they may not honor the agreed-upon decryption key or delete the stolen data. However, financial loss continues even when a ransom is paid because new attacks occur, the decryptors do not work, or the attackers retract their promises. In turn, it will be independent victims who may have to pay multimillion-dollar recovery expenses along with financing criminal activities.

Companies also risk getting entangled in legal troubles related to data protection, anti-money laundering, sanction, or anti-terrorism funding laws when making ransomware payments [26]. The identity of attackers is often unknown, which means that a payment intended to be made for the return of data can instead provide funds for terrorists or other threats to national security. The cost impact on an organization's reputation and public goodwill in cases of high-profile ransomware attacks is also worth noting. Reporting payments to criminal organizations, fines for sanctions breaches, or data leaks might devastate the reputation, especially for companies operating in industries with numerous customers' sensitivities, such as finance or healthcare.

CHAPTER 13   NEGOTIATING WITH ATTACKERS

Finally, it is up to the leadership of an organization to decide whether they are willing to take that certain risk and to determine the best way to preserve the organization for the long term while considering all the visible and hidden consequences. Admissible business continuity and data backup levels could be sufficient to manage these risks to a level that does not leave the company in a compromising position when negotiating contracts. Sophisticatedness and speed in addressing or containing cyber-threats can cause time blindness or buy time. While challenging, one has to avoid the lure of the negotiate-down approach as it starves the ransomware business model of the lifeblood: cash payments. But again, organizations must know they can take that short-term operational hit. Assessment factors include the criticality of the data that is affected, the ransom demanded concerning the probable recovery costs, the legal necessitating reporting of ransomware incidents, the insurance coverage offered, and the total organizational resiliency. There is no perfect script for moving through this paradox, but a strategic organization's consequences on legal/illegal, moral/immoral, financial/chaos costs/benefits, and functional/dysfunctional physical systems, both organizationally and environmentally, are imperative. Though a well-balanced approach to risk-taking and cybersecurity investments can potentially benefit an organization or a business, the essence of proving the hypothesis of the potential superiority of such a decisive process amid the array of variants that can be considered imperfect remains arguable or questionable to a certain extent.

Factors to Consider Before Making Negotiations

In deciding how to counter the effect of ransomware, some aspects would need to be considered first. These will assist in understanding whether the possible gains of paying the ransom outweigh the costs, risks, and other drawbacks of dealing with the hackers.

CHAPTER 13   NEGOTIATING WITH ATTACKERS

Data Criticality

Among the main issues, it is possible to note the type and sensitivity of the information that the attackers have encrypted. Where specific data is essential for carrying out fundamental organizational business or contains particularly private data relating to the client, it increases pressure to get a ransom as a possible option. The gains on the other end might not offset the costs and risks if the downtime lasts long and becomes exponentially More costly than the ransom demand. Still, it is impossible always to ensure that paying back will lead to the recovery of usable data or that the data that has been exfiltrated will not be leaked.

Ransom Payment Feasibility

The very fact that the amount of ransom demanded in dollars is so massive cannot be excluded as playing a definite role. If the ransom ask is absurdly high, it may be impossible to pay, even if circumstances dictate that the data has to be recovered. For instance, some ransoms go as high as the hundreds of thousands, if not reaching the millions, and sometimes even the tens of millions, figures that are unattainable to many victims [27]. Hence, organizations need to be objective about where they can mobilize such a sum of money, and depending on what the funder looks at next, there could be legal problems down that route. Paying may create follow-up demands for even more money as well.

Data Recovery Alternatives

If the victim organization has trustworthy backups of the affected information and systems that can be recovered in a couple of days, this might render the settlement of threat actors completely optional [28]. Any perceived need to bargain with criminals is lessened when remedies are available. Nevertheless, it's frequently easier said than done to restore from backups. Depending on which systems are unavailable, it needs a lot of bandwidth, knowledgeable workers, and a recovery period that might still seriously impact business operations in the interim.

Operational Disruption

The possibility of normal business operations being significantly restricted during the ransomware attack and the organization's capability to recuperate without paying could either draw them closer to or push them away from paying the ransom. If the targeted attack only concerns or affects the less significant or other parts of the organization, then paying a ransom would not be advisable, no matter how big it is. On the other hand, if the organization is threatened with potential bankruptcy due to continued disruption from a severe event, paying may be the best of the two evils to continue the organization's existence; however, legal and moral pitfalls remain.

Third-Party Consequences

Ransomware attacks are now closely linked to data theft; hence, customer records, IP, monetary information, and much more are willing to fall into the wrong hands of criminals. This increases the pressure to one-up this new extortion vector. Companies might pressure themselves to pay ransoms for their data and avoid leakage of any third-party stakeholder's data they manage. Legal consequences such as fines and legal proceedings may result from scenarios involving reckless data handling in the organization under their supervision.

Insurance Considerations

For organizations with a subscription to cyber insurance, it may be logical to report a loss, sit back, and let the insurer take the next step of working with cybercriminals for payment negotiation. Nevertheless, many policies are shifting or are heavily restricted to not covering ransomware events or paying for the losses from cyber extortion. Coverage terms should be analyzed more severely.

Public Relations and Brand Equity

The business implications and overall organizational damage that ransomware can cause, especially to firms' public relations and brand equity, are something that organizational leaders need to consider [29]. B2C organizations and public authorities are most exposed to reputational risks if the systems are breached or customers' data is compromised. The decision relating to negotiation might affect factors such as goodwill, trust, and future sales. Making well-informed decisions about whether and how to respond to ransom demands will be made possible by methodically evaluating all these interconnected aspects. However, there are only the most practical approaches for the specific situation—perfect answers are unattainable. Paying the ransom could have long-term effects on an organization, such as empowering attackers and sustaining the ransomware ecosystem.

Seeking Legal Counsel

Ransomware payment entails legal consequences and legal jeopardy, and it involves numerous factors that may be unique to the country, state, or institution involved. Due to these factors, it would be inadvisable for an organization to handle ransomware attacks or attempt to negotiate with the attackers without first seeking legal and expert advice [30]. Ransomware payments trigger various aspects of laws and regulations, such as the demands for Data Privacy/Protection, Sanctions, Anti-money laundering and Terrorism Financing, Cybercrime, etc. [14, 16]. Due to

CHAPTER 13 NEGOTIATING WITH ATTACKERS

the precise characteristics of each case, organizational leaders will be able to identify the specific legal requirements that regulate this type of activity, as well as possible risks and liabilities. One needs professional legal opinions to evaluate an organization's risk in this complex network fully. Cybersecurity attorneys have been exposed to initial news reports that can reveal details and possible explanations such as the identity of the organization's capabilities meted or stolen, the kind of systems that were compromised, the attacker's money launderers, the geographical coverage of the attack infrastructure, and many more aspects. This indicates that if sanctions are violated, participation in unlawful activities such as support for criminal activity or aiding and abetting liability can flow from paying or refusing to pay a ransom demand.

Also, legal advice brings strategic benefits to eliminating ransomware and the organizational recommendations for a course of action. They can compare the benefits and risks of various strategies, negotiate legal perspectives, preserve any possible legal requirements when cooperating with law enforcement, and put adequate structures in place to meet the standards for possibly saving legal compliance. Data breach disclosure laws and contractual obligations for information security are yet another legal hotspot that cannot be safely navigated without legal assistance [23]. Involving reckless data handling and being cautious in their communication because making statements that could lead organizations to admit liability may lead to lawsuits or regulatory actions regarding the event. Legal advisors can most easily develop these distinct notification patterns, and response postures can also be adopted. Ransomware violations can have dire consequences, including financial penalties, legal action from the government, civil lawsuits, and even criminal prosecution per anti-cybercrime legislation. Ignoring legal counsel is an intolerable risk that threatens the entire company. Their advice is essential for handling ransomware situations properly and complying with all relevant legal requirements.

CHAPTER 13   NEGOTIATING WITH ATTACKERS

Working with Law Enforcement

Legal authorities and law enforcement agencies should be treated as an essential asset when ransom developers attack an organization. This is because reporting the attack is a way through which authorities can apprehend the attackers and make them face the law. It can also offer the authorities valid data for blocking further attacks.

Benefits of Collaboration

Working with cybercrime law enforcement has advantages for organizations. It helps organizations understand how the attack began and its execution. Police authorities and other investigative and counterterrorism agencies possess practical tools and valuable knowledge in solving ransomware cases and attacks [32]. These are forensic activities, threat intelligence, and providing services to the victims of cybercrime. Cooperation with law enforcement entities can help organizations recover their stolen data and minimize the harm done by the attack. Law enforcement bodies like Interpol, the UK National Crime Agency (NCA), Europol, and the FBI can pursue the culprits and get their properties, including the ransoms the victims pay [33]. They are likely to also assist organizations in recovering their codes or securing their systems through copies. Besides, cooperation with them enables organizations to prevent future attacks. Through publicizing an attack, organizations assist related law enforcers in arresting the persons responsible for the attack [33]. This may steer other cybercriminals away from attacking that organization in the future. Law enforcement agencies can also enlighten organizations on how they can improve the next time they are attacked.

CHAPTER 13   NEGOTIATING WITH ATTACKERS

Logistics of Collaboration

It is always best to enter into a working relationship with the police or investigative force in case of a ransomware attack. This should include, for instance, allocating someone from within an organization who can interact with the investigative team and offer them all the required information. Generally, this person should have adequate knowledge and prior experience in the organization's policies regarding cybersecurity and should have the necessary power when making decisions on behalf of the organization. Organizations must also be ready to make data and logs available to law enforcement. All cons regarding the attack, the ransom, delivery of the ransom demand, and any response to the attackers should be documented. One must immediately secure this evidence because it will be helpful in the investigation. It also means that patience and cooperation are essential for organizations anyway. Crime investigations may require a while, and it will be prudent for law enforcement to conduct their work as they wish. Through organizations, persons should also be ready to respond to various questions from law enforcement, besides offering any other details that might be required by law enforcement.

Building Trust and Communication

Trust and the main communication channels between the organizations and the law enforcement departments should exist. Such a relationship means that the organization should be open and provide information, while at the same time, such information should be politely demanded by law enforcement agencies [35]. It entails guarding organizational information and preventing interference with organizational activity. The importance of proven and stable communication comes in since, in this way, both parties are assured that they are in touch and trust each other. Law enforcement involvement requires organizations to keep giving updates on the progress of the attack and recent occurrences that are

CHAPTER 13   NEGOTIATING WITH ATTACKERS

useful in investigations. Law enforcement agencies must also provide such organizations with information concerning the progress of an investigation and any steps taken.

Challenges in Collaboration

As this chapter has shown, there are various rewards to partnering with organizations and law enforcement, but this comes with some risks. However, this relationship is characterized by one main weakness: a lack of trust between the two parties. Of course, organizations are often reluctant to provide information to law enforcement agencies because they may use it to capture an offender or appear in the media. On the part of law enforcement agencies, they may feel agitated by organizations that do not collaborate or provide essential information or by organizations that attempt to influence the investigation process. The other challenge is that most organizations need to be more well resourced. Police departments and other law enforcement. Weement agencies often need to be better equipped and supported by adequate personnel and financial resources to conduct an adequate investigation and respond to cybercrime cases [36]. This can cause investigation time to be dragged and agencies as well as law enforcement to be frustrated. However, there is a need for organizations and law enforcement to come together to fight ransomware as well as other cyber threats. In this way, they can exchange information, services, and knowledge to investigate and punish criminals and restore stolen information and information in future incidents.

Summary

While weighing the legal and ethical aspects of negotiating with ransomware attackers to regain control of the affected systems is essential, there are other practical considerations. Unfortunately, there is no

CHAPTER 13　NEGOTIATING WITH ATTACKERS

straightforward answer to this question, as strategies that should be taken after an attack crucially depend on the particularities of the attack and the organization that has become a victim. Ransomware attacks are increasing daily in different parts of the world, targeting organizations, governments, and people. The implications of these attacks are quite dear, estimated to be in the region of several billion dollars every year. The ransomware attacks are not novel; the world has been feeling the effects of this terror for quite some time now, including in the United States, Europe, and others. It is unlawful and unethical to pay a ransomware attacker, as it will attract more attacks in the future. Giving to law enforcement may be against the anti-terrorism laws since it involves paying for criminal activity, which may lead to more attacks. Conversely, failing to pay will likely lead to a permanent loss of files and disruption of the organization's operation. In general, negotiation is a crucial element in accomplishing the goals and objectives of organizations. However, various risks should be taken into consideration before an organization decides to negotiate. There are specific criteria one should keep in mind while considering the negotiation with the ransomware attackers, such as the criticality of the encoded data, the chances of paying the ransom, the availability of the data recovery option, current and potential operational disturbance impact, third party concern, insurance coverage, and brand and image issues. You should get assistance from an attorney when facing ransomware attacks. Such legal counsel comprises attorneys advising an organization on the legal repercussions of paying ransom, clarifying the organization's rights and wrongs, and formulating resourceful ways of containing the attack. Another factor involves collaboration with law enforcement, which is also crucial. Informing the police about ransomware attacks can help the police arrest those behind the vice, recover stolen data, and prevent the reoccurrence of heinous crimes. Close cooperation should be between the organizations and the police to address the increasing ransomware problem. That is why it is reasonable to state that negotiating with ransomware attackers is quite a challenging problem without any universal and rather simple investigation process. We must consider several factors before deciding its legal, ethical, and practical ramifications. Working with law enforcement and obtaining legal representation are crucial measures for effectively addressing ransomware threats.

References

[1] ProofPoint, "Ransomware—Definition, Prevent Attacks & Viruses | Proofpoint US," *Proofpoint*, Dec. 23, 2020. https://www.proofpoint.com/us/threat-reference/ransomware#:~:text=Ransomware%20is%20a%20type%20of

[2] K. Yasar, "What is Disaster Recovery (DR)?," *SearchDisasterRecovery*, 2022. https://www.techtarget.com/searchdisasterrecovery/definition/disaster-recovery

[3] Dmarcian, "FBI Releases 2020 Internet Crime Report," *dmarcian*, Mar. 25, 2021. https://dmarcian.com/fbi-2020-internet-crime-report/ (accessed May 31, 2024).

[4] eSentire, "Cybercrime To Cost The World $9.5 Trillion USD Annually In 2024," *eSentire*, 2024. https://www.esentire.com/web-native-pages/cybercrime-to-cost-the-world-9-5-trillion-usd-annually-in-2024#:~:text=Ransomware%20will%20cost%20its%20victims

[5] Enisa, "Threat Landscape," *ENISA*, 2022. https://www.enisa.europa.eu/topics/cyber-threats/threats-and-trends

[6] Cloudflare, "What was the WannaCry ransomware attack?" *Cloudflare*, 2BC. https://www.cloudflare.com/learning/security/ransomware/wannacry-ransomware/#:~:text=On%20May%2012%2C%202017%2C%20the,its%20ambulances%20to%20alternate%20hospitals

[7] G. Harding, "WannaCry Ransomware attack costs £92 million," *IT Governance UK Blog*, Oct. 12, 2018. https://www.itgovernance.co.uk/blog/wannacry-ransomware-attack-costs-92-million

[8] E. M. Lab, "The cost of ransomware in 2021: A country-by-country analysis," *Emsisoft | Cybersecurity Blog*, Apr. 27, 2021. https://www.emsisoft.com/en/blog/38426/the-cost-of-ransomware-in-2021-a-country-by-country-analysis/

[9] A. Din, "New Report Shows Global Ransomware Attempts Volume Hits 304.7 Million in 2021," *Heimdal Security Blog*, Aug. 02, 2021. https://heimdalsecurity.com/blog/new-report-shows-global-ransomware-volume-reached-304-7-million-attempted-attacks-in-the-1st-half-of-2021/ (accessed May 31, 2024).

[10] Singapore Police, "FAQs on Ransomware," *Singapore Police Force*. https://www.police.gov.sg/Advisories/Crime/Cybercrime/Ransomware/FAQs-on-Ransomware#:~:text=There%20is%20no%20guarantee%20that (accessed May 31, 2024).

[11] Acronis, "The legal implications of paying ransomware demands: The evolving state of ransomware," *Acronis*, 2022. https://www.acronis.com/en-us/blog/posts/the-legal-implications-of-paying-ransomware-demands-the-evolving-state-of-ransomware/

[12] A. M. Freed, "What are the Legal Implications from a Ransomware Attack?," www.cybereason.com. https://www.cybereason.com/blog/what-are-the-legal-implications-from-a-ransomware-attack

[13] C. Harrigan, "Negotiating with Ransomware Attackers: Ethical and Legal Considerations," *VENZA*®, Jan. 22, 2024. https://www.venzagroup.com/negotiating-with-ransomware-attackers-ethical-and-legal-considerations/

[14] U.S. Department of the Treasury, "United States Sanctions Affiliates of Russia-Based LockBit Ransomware Group," *U.S. Department of the Treasury*, Feb. 20, 2024. https://home.treasury.gov/news/press-releases/jy2114

[15] American Hospital Association, "HC3 Intelligence Briefing Dridex Malware," 2020. Available: https://www.aha.org/system/files/media/file/2020/06/hc3-cyber-threat-briefing-tlp-white-dridex%20malware-6-25-2020.pdf

[16] U.S. Department of the Treasury, "Treasury Sanctions Evil Corp, the Russia-Based Cybercriminal Group Behind Dridex Malware | U.S. Department of the Treasury," home.treasury.gov, Dec. 05, 2019. https://home.treasury.gov/news/press-releases/sm845

[17] "What is Ransomware? | KnowBe4," www.knowbe4.com. https://www.knowbe4.com/ransomware#:~:text=It%20does%20not%20delete%20any (accessed May 31, 2024).

[18] Homeland Security, "Ransomware Attacks on Critical Infrastructure Sectors." Available: https://www.dhs.gov/sites/default/files/2022-09/Ransomware%20Attacks%20.pdf

[19] E. A. Feigenbaum and M. R. Nelson, "Korea's Path to Digital Leadership: How Seoul Can Lead on Standards and Standardization," Feb. 16, 2024. https://carnegieendowment.org/research/2024/02/koreas-path-to-digital-leadership-how-seoul-can-lead-on-standards-and-standardization?lang=en

[20] W. Rodriguez, "10 Biggest Ransomware Attacks | Blog | Fluid Attacks," fluidattacks.com. https://fluidattacks.com/blog/10-biggest-ransomware-attacks/#:~:text=REvil%20also%20carried%20out%20another (accessed May 31, 2024).

[21] G. Geyer, "Cyber Attack Overview: JBS Foods Ransomware Incident," *Claroty*, 2023. https://claroty.com/blog/jbs-attack-puts-food-and-beverage-cybersecurity-to-the-test#:~:text=The%20Brazilian%20food%20company%20and (accessed May 31, 2024).

[22] A. Waldman, "JBS USA paid $11M ransom to REvil hackers I TechTarget," *Security*, Jun. 2021. https://www.techtarget.com/searchsecurity/news/252502288/JBS-USA-paid-11-million-ransom-to-REvil-hackers (accessed May 31, 2024).

[23] T. Hofmann, "How organizations can ethically negotiate ransomware payments," *Network Security*, vol. 2020, no. 10, pp. 13–17, Oct. 2020, doi: https://doi.org/10.1016/s1353-4858(20)30118-5

[24] P. Sumner and J. Simons, "Ethical Legal Implications of Paying Ransoms," www.nationaldefensemagazine.org, Aug. 17, 2021. https://www.nationaldefensemagazine.org/articles/2021/8/17/ethical-legal-implications-of-paying-ransoms

[25] N. Gaur, "Held hostage by ransomware? Here's how to respond," *World Economic Forum*, Aug. 19, 2019. https://www.weforum.org/agenda/2019/08/held-hostage-by-ransomware-heres-how-to-respond/ (accessed May 31, 2024).

[26] International Monetary Fund, "Anti-Money Laundering and Combating the Financing of Terrorism," *IMF*, 2023. https://www.imf.org/en/Topics/Financial-Integrity/amlcft

[27] J. Konetschni, "Council Post: Eight Steps To Negotiating With Ransomware Hackers," *Forbes*, 2023. https://www.forbes.com/sites/forbesbusinesscouncil/2023/07/20/eight-steps-to-negotiating-with-ransomware-hackers/?sh=626eb97b5b01 (accessed May 31, 2024).

[28] C. C. for C. Security, "Tips for backing up your information (ITSAP.40.002)," *Canadian Centre for Cyber Security*, Sep. 28, 2020. https://www.cyber.gc.ca/en/guidance/tips-backing-your-information-itsap40002

[29] J. Maccoll *et al.*, "The Scourge of Ransomware Victim Insights on Harms to Individuals, Organisations and Society," 2024. Available: https://static.rusi.org/ransomware-harms-op-january-2024.pdf

[30] W. Gotshal, M. L.-J. Maples, H. Lund, R. Meehan, and S. Chaplin, "Falling Victim to a Ransomware Attack - Key Legal Considerations," *Lexology*, May 14, 2024. https://www.lexology.com/library/detail.aspx?g=26f3e21b-d1f7-45fc-83f1-fb00c28edca0#:~:text=In%20this%20rapidly%20evolving%20area (accessed May 31, 2024).

[31] J. Ledesma, "The Risks and Legal Implications of Failing to Disclose a Security Breach," *Bitdefender Blog*, 2023. https://www.bitdefender.com/blog/businessinsights/the-risks-and-legal-implications-of-failing-to-disclose-a-security-breach/

[32] D. Burns, B. Williamson, G. Dunn, and amp; Crutcher', "First published on the Global Investigations," 2021. Available: https://www.gibsondunn.com/wp-content/uploads/2021/10/Burns-Williamson-Should-companies-cooperate-with-law-enforcement-during-ransomware-attacks-GIR-10-08-2021.pdf

[33] Application Security Series , "TOP 10 Law Enforcement Agencies Most Active in Fighting Cybercrime," www.immuniweb.com, 2022. https://www.immuniweb.com/blog/top-10-law-enforcement-agencies-cybercrime-fraud.html

[34] Atos, "How government and law enforcement can help prevent ransomware attacks," *Atos*, 2024. https://atos.net/en/lp/turning-tables-on-ransomware/how-government-and-law-enforcement-can-help-prevent-ransomware-attacks

[35] A. Mandelbaum, "Ethical Communication: The Basic Principles," *Paradox Marketing*, Oct. 23, 2020. https://paradoxmarketing.io/capabilities/knowledge-management/insights/ethical-communication-the-basic-principles/

[36] J. Curtis and G. Oxburgh, "Understanding Cybercrime in 'Real World' Policing and Law Enforcement," *The Police Journal: Theory, Practice and Principles*, vol. 96, no. 4, Jun. 2022, doi: https://doi.org/10.1177/0032258x221107584

[37] EMSISOFT MALWARE LAB, "The State of Ransomware in the US: Report and Statistics 2022," *Emsisoft | Cybersecurity Blog*, Jan. 02, 2023. https://www.emsisoft.com/en/blog/43258/the-state-of-ransomware-in-the-us-report-and-statistics-2022/

CHAPTER 14

Rebuilding and Strengthening Security Posture

Author:
Anirudh Khanna

Cyberattack-Learning from the Incidents

The escalating frequency of cyberattacks and data breaches is a pressing concern for individuals and organizations alike. As of April 2024, the staggering number of over 4 billion records breached globally underscores the urgency of the situation. Shockingly, only 43% of these incidents were reported, indicating a significant underestimation of the problem. The United States, in particular, is a hotspot for data breaches, accounting for over 80% of global incidents. The following table provides a sector-wise overview of the top 5 most breached sectors, further emphasizing the need for immediate action.

CHAPTER 14   REBUILDING AND STRENGTHENING SECURITY POSTURE

Table 14-1. Top 5 Most Breached Sectors [6]

Sector	Cyberattack Incidents	
Healthcare	81	29%
I.T. services	29	10%
Manufacturing	24	9%
Education	20	7%
Finance	18	6%

In Table 14-1, the healthcare sector experienced the highest number of cyberattacks in April 2024. The second was I.T. services, with 10%; third was manufacturing services, with 9%; fourth was the education sector, with 7%; and lastly, the finance sector, with 6%. However, these are only publicly disclosed incidents. Other undisclosed incidents protect the organization's image.

Table 14-2. Top 5 Breached Sectors by Number of Records [6]

Sector	Records breached
I.T. services and software	4 billion
Insurance	15million
Healthcare	5 million
Manufacturing	4 million

In Table 14-2, the I.T. services and software sector experienced the highest number of records breached by April 2024. The second was the insurance sector, with 15 million. The healthcare and manufacturing sectors had 5 million and 4 million, respectively.

As illustrated in the figure, the number of monthly disclosed incidents reached a record high in March 2024,

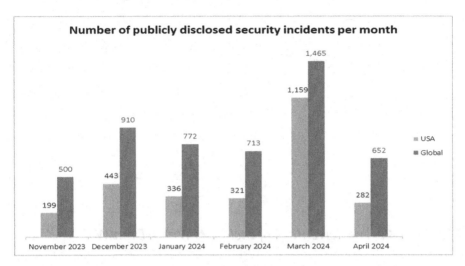

Figure 14-1. Number of Publicly Disclosed Security Incidents Per Month [6]

In Figure 14-1, March 2024 had the highest security incidents. Globally, there were 1,465 incidents, while in the United States, there were 1,159 incidents. November 2023 recorded the lowest number of incidents, with 199 incidents in the United States and 500 globally.

Regarding data breaches, the United States still topped the percentage of incidents in which data was breached in various sectors.

CHAPTER 14   REBUILDING AND STRENGTHENING SECURITY POSTURE

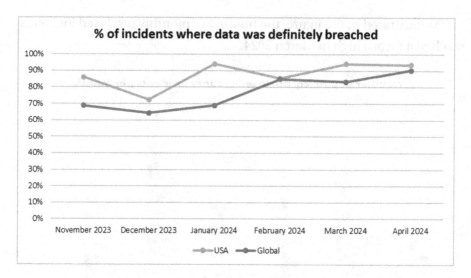

Figure 14-2. The Percentage of Incidents Where Data Was Breached [6]

In Figure 14-2, the United States experienced over 90% of incidents where data was breached. The highest global percentage of incidents was in January 2024, with 87%. Figure 14-2 shows that the incidents where data was breached have increased since 2023. In 2024, there are still incidents of data breaches, thus raising concerns about the strength of the security features in various sectors of the world, starting with the United States. Moreover, these incidents are the only ones publicly disclosed. Other many incidents have not been publicly disclosed.

Some organizations have taken action in response to cyberattack incidents. The following figure illustrates.

CHAPTER 14 REBUILDING AND STRENGTHENING SECURITY POSTURE

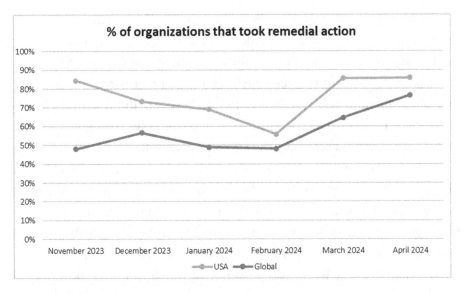

Figure 14-3. Percentage of Organizations That Took Action [6]

In Figure 14-3, the organizations in the United States took the highest number of remedial actions, especially in November 2023, March 2023, and April 2023, when they took remedial action for more than 80% of the incidents. Conversely, global action to the incidents was low, starting at 47% in November 2023, dropping between December and February 2023, and rising steadily in March and April 2024.

Learning from past incidents is crucial for continuous improvement in cybersecurity. The types of attacks recorded include social engineering, ransomware, supply chain, and brute force attacks. It is essential to conduct a post-review to learn from the attack because there are different ways an adversary can disrupt the operations of a software system. From the discussion on the number of incidents, the I.T. sector and software recorded the highest number of data breaches (4 billion) by March 2024 [5]. Therefore, it is essential to understand how these data breaches occur to make an informed decision when closing the loopholes.

CHAPTER 14   REBUILDING AND STRENGTHENING SECURITY POSTURE

Social engineering refers to the attempt by attackers to manipulate users into clicking on malicious links. The adversary gains access to the system once a user clicks on the provided link or downloads an attachment. In a brute force attack, an attacker consistently uses different user names and passwords to gain access to the system [2]. The developers, therefore, should ensure the passwords set are highly complex and that the chances of getting the correct combination are almost zero.

Ransomware attacks encrypt the system's information, thus locking out the user. Once inside the system, the attacker can lock the available data until a ransom is paid. Ransomware attacks are expensive to individuals and organizations because they have to decide whether to pay the ransom or use other means to recover data. Even when the victims choose to recover data, the cost is still as high as paying the ransom. The primary motivation of ransomware attackers is to get a hefty payout. The following figure illustrates the number of disclosed ransomware attacks in the world.

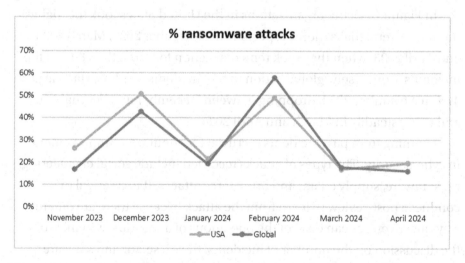

Figure 14-4. *Percentage of Ransomware Attacks. Source: [7]*

In Figure 14-4, the United States experienced the highest number of ransomware attacks, with 50% occurring in December 2023 and 68% occurring in February 2024. March 2024 recorded the lowest number of ransomware attacks, with less than 20% for the United States and global attacks.

Lastly, supply chain attacks occur when an attacker injects malicious code into a product or a software update. Supply chain attacks aim at disrupting or diverting the supply chain to the location of their choice. The National Institute of Standards and Technology states that cyberattackers can access the organization's network using a compromised system. The attackers can also bypass the security features to access a system. For example, in the SolarWinds and Kaseya breaches, the attackers bypassed security measures and caused significant damage to the organization's resources, thus damaging and disrupting the distribution networks [3].

Therefore, it is essential to conduct a post-incident review to learn from the attack and improve future security posture. The organization knows what went wrong and how to fix the issues presented by performing a post-incident review. Moreover, a post-incident review evaluates security risks and the features an organization must implement to prevent future attacks.

Implementing Security Enhancements

The post-recovery features should include implementing security enhancements that will prevent future attacks. This includes technological upgrades, policy adjustments, and ongoing staff training. Most importantly, an organization must adopt a layered approach to cybersecurity, which involves monitoring across networks and endpoints. A layered security approach means that each feature or tool on a network offers additional security such that the failure of one feature will not break the system's defense. For instance, suppose a user clicks on a malicious link, an antivirus should block the attack.

Policy Adjustments

A cyberattack that causes significant damage to an organization indicates that the organization needs to adjust its policies regarding cybersecurity. For instance, in 2021, the Houston-based Colonial Pipeline was attacked by ransomware, disrupting the energy supply to millions of people [3]. The ransomware primarily targeted the company's Information Technology (I.T.) systems. However, the company shut down the pipeline to prevent further entry into the I.T. system, which could have exposed confidential information. Ultimately, Colonial Pipeline spent millions to contain the attack and recover the damage. The Colonial Pipeline case is one of the many cases that occur worldwide. Some are disclosed, while some companies pay ransom and keep the matter confidential to maintain the trust of stakeholders and shareholders.

However, companies like Colonial Pipe need to adjust policies in post-attack recovery. These policies will help the organization strengthen its security posture instead of waiting for the attack to occur before making adjustments. These policies are related to the cybersecurity framework to adopt.

Cybersecurity Framework

The cybersecurity framework refers to the guidelines and standards for risk management in cybersecurity. If an adversary succeeds in attacking an organization's I.T. system, it means loopholes must be addressed [10]. Adopting a cybersecurity framework will help to identify and close the loopholes. Research shows that adopting a cybersecurity framework decreases cybersecurity risk by up to 85%. There are several existing cybersecurity frameworks, such as the National Institute of Standards and Technology (NIST), Service Organization Control (SOC), ISO Certification, and CIS Controls, among others [4]. These frameworks are

CHAPTER 14 REBUILDING AND STRENGTHENING SECURITY POSTURE

equally important depending on the organization type and the scope of the cyberattack [10]. For instance, an organization likely to lose or spend millions of dollars on recovery must adopt a cybersecurity framework that addresses its needs effectively.

To provide holistic security, an organization post-recovery should adopt CIS controls. Moreover, user reviews show that CIS controls are relatively inexpensive and user-friendly. As stated in [8], CIS controls defend against common cyberattacks by reducing the attack surface and mitigating risks such as data breaches and extended downtimes. CIS controls are also standardized and provide consistent protection against potential cyberattacks [2]. Moreover, the processes are easy to follow regardless of the organization's size. Therefore, an organization in the post-recovery phase should consider implementing the CIS controls. Research shows that using the CIS Controls framework reduces the risk of cyberattacks by up to 85%. If the organization already adopted another framework, the management should have a policy to review the frameworks regularly to ensure they meet the organization's current needs [5].

How can an organization know a practical cybersecurity framework to adopt post-recovery? An effective framework provides a guide for determining risks and opportunities for addressing the risks to strengthen control. Also, a helpful framework considers and provides opportunities for continuously monitoring the I.T. system to detect and prevent future cyberattacks. Moreover, a practical cybersecurity framework directs the organization on where to invest the resources to provide holistic security to the I.T. systems [3]. Overall, cybersecurity frameworks are part of governments' policy requirements to mitigate cybersecurity attacks. However, institutions must adopt cybersecurity frameworks such as CIS Control, not because it is a requirement by the government, but to prevent future cyberattacks.

CHAPTER 14   REBUILDING AND STRENGTHENING SECURITY POSTURE

Create a Cybersecurity-Centric Culture

A post-recovery organization should embrace a cybersecurity-centric culture through continuous cybersecurity training and awareness. Staff training should continue as planned, but the organization should adopt a culture that fosters cybersecurity awareness. Organizations can outsource aspects of cybersecurity, such as equipment and human resources, but promoting a cybersecurity-centric culture is different [2]. For a post-recovery organization, employees are the first line of defense against cyberattacks, even when the most sophisticated technology is available. Most cyberattacks occur because some people knowingly or unknowingly fail to perform their duties.

If employees work remotely, a cybersecurity-centric culture becomes more critical because the organization must defend the extended network of people against cyberthreats. An organization's first step is to instill a cybersecurity-focused mindset [1] [2]. The employees should know they are the institution's first defense against attacks. Employees can perform their duties in various ways. For example, creating strong passwords on their devices, including phones, home routers, and personal computers [5]. The company can adopt sophisticated systems to enhance cybersecurity post-recovery. However, with the failure of the employees to comply and act as the first line of defense, attackers can still bypass this cybersecurity system. Therefore, the management should create a culture of cybersecurity by showing the employees how protecting the organization's I.T. system is aligned with their personal goals. In this way, employees can see what is in it for them to create a cybersecurity-centric culture.

In the post-recovery phase, the company should also implement continuous cybersecurity training programs to educate the employees about the latest cyberthreats and the best practices to safeguard their devices and the company's equipment against attackers. Keeping employees on guard against cyberthreats, especially regarding the social engineering schemes that attackers use to obtain usernames and passwords, will improve cybersecurity. Adopting CIS controls decreases the number of cyberthreats by up to 85%. Other mechanisms, such as a cybersecurity-centric culture, among other cybersecurity measures, can reduce the remaining 15%.

Employees' failure in social engineering tests means the organization is still at a high risk of cybersecurity attacks. Suppose the employees fail and remain vulnerable to the phishing strategies of potential attackers. In that case, the organization should consider re-educating them on social engineering techniques and how to guard themselves against knowingly or unknowingly providing confidential information to potential attackers [6]. For example, employees should learn not to click on any link that requests them to click on it and obtain more information. The trainers should create a strategy that motivates employees to think about cybersecurity first because the safety of the I.T. systems starts with them. Management should also provide incentives for employees to follow cybersecurity guidelines constantly. Lastly, the organization's leadership should assign cybersecurity ambassador roles to the employees who show effort in championing the security of the company's I.T. systems.

Technological Upgrades

Technological upgrades refer to updating the available technologies to handle cybersecurity threats. After a review and containment of a cyberattack, an organization learns about the cause of the attack and the entry point. An adversary can use various weak links to implement a cyberattack, including outdated technology [2]. An obsolete technology poses risks of attacks and costly data restoration during data recovery because it may need to back up the resources required. Therefore, an organization that is post-recovery should adopt or upgrade the existing technology to back up data regularly.

CHAPTER 14   REBUILDING AND STRENGTHENING SECURITY POSTURE

Technological upgrades help not only prevent cyberattacks but also help recover from them. A post-recovery organization should consider technological upgrades because, in a ransomware attack, the organization can recover data from the backups. This means the system should be able to carry out consistent and regular backups. Ransomware succeeds by holding data captive until the victim pays the requested ransom. Suppose data has been backed up somewhere else. In that case, the attacker becomes less threatening because if the ransom is not paid, the company can obtain data from the backup, upgrade security measures, and continue operations. The only disadvantage will be the sensitivity of the data held by a hacker, which may damage the company's reputation if exposed [4].

Therefore, organizations in post-recovery should install segregated backups to ensure that an attacker does not capture and lock all the existing backups. Suppose an institution has online backups only, and the adversary seizes the network; the backups become useless. Institutions must implement robust access controls and conduct tests on backups to ensure they can be restored successfully in case of a cyberattack [2]. Institutions in post-recovery should consider the following questions when implementing the backup strategy:

- How regularly are the backups being upgraded and tested?
- What are the most essential features of the backup that must be included in the backup strategy?
- Are there any gaps in the backup? If yes, how are they being addressed?
- Where are the backups located? Are they centralized or segregated?

Answering these questions will help an organization in post-recovery to build a strong and resilient backup that can rely on in the event of another attack. For instance, the backup should be tested and upgraded

CHAPTER 14   REBUILDING AND STRENGTHENING SECURITY POSTURE

regularly to close the loopholes identified during testing. The most essential features of the backup include the ability to recover data, protect data against ransomware attacks, and scalability. If gaps are present in the backup, they should be promptly addressed–this means the I.T. team should conduct constant system monitoring. Continuous system monitoring can be achieved by adopting blockchain technology. Lastly, the backups must be segregated to prevent a cyberattacker from accessing all the available backups.

Monitoring and Managing Threats Real-Time

It would be paramount for an institution to implement accurate time monitoring and management of the I.T. systems if a threat bypasses prevention mechanisms in the first place. One of the tools that provides real-time monitoring and management of I.T. systems is the Security Information and Event Management (SIEM) [1] [2]. SIEM detects cybersecurity threats and provides real-time information about cyberattacks, including system vulnerabilities. The main advantage of SIEM is the provision of holistic reviews of the events on an I.T. system [2]. The disadvantage of SIEM, however, is that it is expensive, and its configuration requires time and other resources–this implies that organizations that purchase SIEMS are less likely to configure it to the required standards because of the cost of additional resources needed. Consequently, the organization needs to check the alerts regularly. The failure to check alerts paves the way for data breaches.

For example, in 2013, Target installed SIEMS but failed to check the alerts from the anti-intrusion resource, thus creating a significant breach. Once the hackers accessed the I.T. system, they obtained the Target employees' names and credit card numbers. After mishandling employee information, Target spent over $200 million in legal fees [7]. For a post-recovery institution, it is essential to consider adopting SIEM-as-a-Service (SIEMaaS) to ensure constant monitoring of the systems and provide real-time updates. SIEMaaS is a third-party software that collects the system's logs and sends them to the parent SIEM, which is usually outsourced. The internal I.T. department has direct access to the alerts.

The vendor's security operations center is responsible for managing the outsourced SIEM; this ensures the internal I.T. team is not burdened by restoring the system. An outsourced SIEM is advantageous to an organization because the third party manages all the alerts received. Therefore, a post-recovery organization should consider adopting a managed security service provider (MSSP) because the burden of monitoring will be on the outsourced company, not the internal I.T. team [2]. Also, MSSP will configure the SIEMs to ensure alerts are received, thus giving an institution advanced monitoring of its I.T. systems.

Penetration Testing and Vulnerability Scanning

After implementing the strategies for strengthening the institution's cybersecurity posture, the organization should now turn to penetration testing and vulnerability scanning. While an organization may configure its security features correctly, the IT team should start thinking like hackers and perform penetration tests. Once the anomalies or weaknesses have been detected in the system, they are promptly fixed [2]. However, penetration testing should not be a timed event; it should be done regularly to ensure the system still operates per the required standards. In addition to internal penetration tests, the institution should adopt external penetration tests to determine the ability of the cybersecurity systems to block a cyberattack [9].

External penetration testing attempts to penetrate the organization's features on a network via the Internet. This is because most of the systems are attacked via the internet. The ethical hacker will find ways to bypass external systems and access confidential information, thus discovering how a cyberattacker could disrupt an organization's systems. The primary aim of penetration tests should be to identify the weaknesses in the system and make recommendations on how they can be fixed. The systems tested include firewalls and perimeter routes. Vulnerability scanning refers to assessing the software system parts that require attention in internal and external tests. A post-recovery organization should only partially focus on performing these tests, reviewing the results, and fixing the detected loopholes [2].

Summary

Post-recovery organizations should focus primarily on creating a cybersecurity-centric culture to ensure employees' behavior is aligned with the organizational goals of enhancing cybersecurity. Penetration testing and vulnerability scanning should be the last step in strengthening cybersecurity posture because they test how software systems will behave during a cyberattack. Therefore, the tests should be performed only after the relevant cybersecurity tools, such as monitoring and managing threats, multifactor authentication, and data backups, have been implemented successfully.

References

[1] Ahmad, Aftab. "A Protocol for Monitoring Network Threats in Real-Time." *2023 Congress in Computer Science, Computer Engineering, & Applied Computing (CSCE)*, July 24, 2023. https://doi.org/10.1109/csce60160.2023.00376.

[2] CSI. "Cybersecurity Best Practices." Cybersecurity Best Practices | Cybersecurity and Infrastructure Security Agency CISA, 2023. https://www.cisa.gov/topics/cybersecurity-best-practices.

[3] CSIS. "Significant Cyber Incidents: Strategic Technologies Program." CSIS, 2024. https://www.csis.org/programs/strategic-technologies-program/significant-cyber-incidents.

[4] Das, Ravi. "Cybersecurity Audits, Frameworks, and Controls." *Assessing and Insuring Cybersecurity Risk*, July 27, 2021, 51-92. https://doi.org/10.1201/9781003023685-2.

[5] Donaldson, Scott E., Stanley G. Siegel, Chris K. Williams, and Abdul Aslam. "Cybersecurity Frameworks." *Enterprise Cybersecurity*, 2015, 297-309. https://doi.org/10.1007/978-1-4302-6083-7_17.

[6] Ford, Neil. "Data Breaches and Cyber Attacks in 2024 in the USA." I.T. Governance USA Blog, May 10, 2024. https://www.itgovernanceusa.com/blog/data-breaches-and-cyber-attacks-in-2024-in-the-usa.

[7] Kosling, Kyna. "Data Breaches and Cyber Attacks in the USA in April 2024—4,277,728,098 Records Breached." I.T. Governance USA Blog, May 9, 2024. https://www.itgovernanceusa.com/blog/data-breaches-and-cyber-attacks-in-the-usa-in-april-2024-4277728098-records-breached.

[8] Luna, Ariel, Yair Levy, Gregory Simco, and Wei Li. "Towards Assessing Organizational Cybersecurity Risks via Remote Workers' Cyberslacking and Their Computer Security Posture." *Proceedings on cybersecurity education, research and practice*, November 14, 2022. https://doi.org/10.32727/28.2023.5.

[9] Moseley, Ralph. "Packet Analysis and Penetration Testing." *Advanced Cybersecurity Technologies*, October 28, 2021, 73–94. https://doi.org/10.1201/9781003096894-5.

[10] Srinivasan, Piya. *Strengthening cybersecurity frameworks remains a Global Challenge*, December 10, 2023. https://doi.org/10.54377/d1ef-9335.

PART V

Real-World Perspectives

Chapter 15: Case Studies in the E-commerce Industry

- **Real-World Examples of Security Successes:** Presents detailed case studies of organizations that successfully defended against or mitigated the effects of ransomware attacks, highlighting effective strategies and lessons learned.

- **Failures and Lessons Learned:** Analyzes cases where organizations were significantly impacted by ransomware, examining the failures in security practices that led to successful attacks and the lessons that can be learned.

Chapter 16: Ransomware, Inc.: The Business and Economics of Digital Extortion

- **Understanding the Ransomware Economy:** Insights into the financial ecosystem of ransomware, exploring its transition from small-scale operations to an organized industry.

- **Role of Cryptocurrency:** Examining how cryptocurrencies facilitate anonymous transactions, driving the growth of ransomware.

- **Economic Impact on Victims:** Discussing direct and indirect costs, including operational disruptions and long-term reputational damage.

- **Profitability and Risk-Return Analysis:** Evaluating the business models of ransomware groups and their strategies for maximizing returns while minimizing risks.

- **Preventive Strategies:** Proposing economic interventions and incentives to deter attackers and reduce ransomware profitability.

Chapter 17: Case Studies in Confidential Computing

- **Overview of Confidential Computing:** Introduction to the concept, technologies, and practices that ensure data is encrypted and protected during processing.

- **Real-World Applications:** Detailed analysis of how organizations across various industries implement confidential computing to secure sensitive data.

- **Challenges and Solutions:** Exploration of common challenges encountered in confidential computing and the strategies used to overcome them.

- **Impact Analysis:** Assessment of how adopting confidential computing affects security posture, compliance, and operational efficiency.

- **Future Outlook:** Speculation on the evolution of confidential computing technologies and their potential future applications.

PART V   REAL-WORLD PERSPECTIVES

Chapter 18: Case Studies in Cloud Computing

- **Introduction to Cloud Security:** Basic concepts of cloud security, including the shared responsibility model and key security measures.

- **Diverse Case Studies:** In-depth examination of different scenarios where cloud computing solutions were deployed, focusing on security strategies and outcomes.

- **Lessons Learned:** Compilation of key takeaways from each case study, providing readers with actionable insights and best practices.

- **Risk Management:** Discussion on identifying, mitigating, and managing risks in a cloud environment.

- **Emerging Trends:** Analysis of new and emerging trends in cloud security technologies and practices.

Chapter 19: Case Studies in Enterprise Security Architecture

- **Foundational Concepts:** Explanation of the components of an enterprise security architecture and their roles in maintaining organizational security.

- **Strategic Implementations:** Case studies showcasing how various enterprises have designed and implemented their security architectures.

- **Challenges and Responses:** Detailed look at common obstacles enterprises face in securing their digital assets and the strategies used to address these challenges.

PART V REAL-WORLD PERSPECTIVES

- **Performance Evaluation:** Evaluation of security architecture performance in real-world conditions, focusing on effectiveness and adaptability.
- **Advancements and Innovations:** Discussion on the evolution of enterprise security architecture and potential future developments.

Chapter 20: Case Studies in Energy

- **Industry-Specific Threats:** Highlighting unique vulnerabilities in the energy sector, such as SCADA systems and critical infrastructure dependencies.
- **Impact of Ransomware:** Analyzing the operational, environmental, and economic consequences of ransomware attacks in energy.
- **Detailed Case Studies:** Reviewing incidents like the Colonial Pipeline attack to illustrate lessons in resilience and recovery.
- **Strategic Responses:** Evaluating effective response measures, including partnerships with government agencies and improved sector-specific frameworks.
- **Future Implications:** Predicting trends in ransomware targeting energy, emphasizing the need for proactive measures.

Chapter 21: Securing Digital Foundations: A Point of View on Cybersecurity in Healthcare

- **Ransomware's Role in Healthcare:** Exploring how ransomware targets healthcare systems, disrupting patient care and operations.

- **Preventative Measures:** Discussing sector-specific safeguards, such as EHR system protection and HIPAA compliance.

- **Recovery Strategies:** Reviewing the unique challenges of restoring healthcare services post-attack, balancing speed and data integrity.

- **Future Trends in Healthcare Security:** Emphasizing the integration of AI-driven solutions and advanced encryption to combat emerging threats.

- **Lessons from Case Studies:** Drawing from real-world scenarios to recommend best practices tailored to healthcare environments.

CHAPTER 15

Case Studies in the E-commerce Industry

Author:
Dhruv Seth

Overview of Ransomware Attacks in the E-commerce Industry

As the e-commerce industry became more significant in 2020 during the COVID-19 pandemic as a method of making sales and purchases, considerable increases in ransomware attacks were recorded. The nature of e-commerce itself makes the industry especially attractive for cybercriminals because of the need by both vendors and buyers to surrender information related to their personal and financial data [1]. Since 2020, when the COVID-19 pandemic forced most global retail commerce to go online, ransomware attacks have spiked across the board, with particular interest in larger, more lucrative companies. The surge in funding for cybersecurity in e-commerce firms between 2020 and 2021, as demonstrated by Figure 15-1, reflects the spike in ransomware and other forms of malware. E-commerce firms spent more than $21 billion in 2021, up from just $8.3 billion the previous year, demonstrating how much mitigating malware like ransomware attacks cost these firms and the industry.

CHAPTER 15   CASE STUDIES IN THE E-COMMERCE INDUSTRY

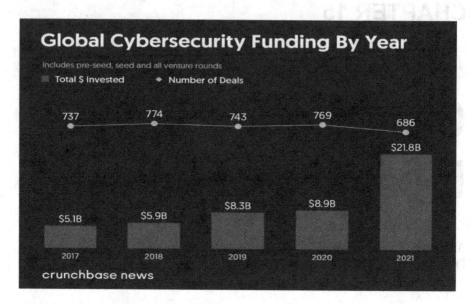

Figure 15-1. Total Spending on Cybersecurity in the E-commerce Space Between 2017 and 2021 [2]

Several considerations make ransomware attacks unique in the e-commerce sector. The first factor is phishing, which entails social engineering personal details from reckless or unsuspecting customers using e-commerce applications on their smart devices or personal computers [3]. The designers of such ransomware create links and place them on existing e-commerce websites or applications, which tricks victims into clicking them, thus planting malware in their devices and starting the ransomware attacks.

The second factor is e-skimming, which entails using brute force attacks or compromised e-commerce websites and applications, exposing the victim's credit card to attacks. Once the customer enters their personal credit card information, the attackers take over its credentials, commencing their ransomware attack by demanding payment from victims or organizations targeted [4]. This ransomware attack is joint for online stores where credit card information must be input for the transaction to end and the purchase to get delivered.

CHAPTER 15 CASE STUDIES IN THE E-COMMERCE INDUSTRY

Third, cross-site scripting (XSS) is an advanced ransomware attack in which unique malware code is injected into e-commerce sites and applications using XSS to target users [5]. Once the users begin using their applications or compromised sites, their login and personal details are recorded discretely, granting the attack access to potentially thousands of users and their personal and financial data.

The final factor in ransomware attacks within the e-commerce industry is SQL injections, which target industry vendors who store transaction and customer data in SQL databases and servers [6]. Once the SQL is injected into the database, it raises unauthorized queries targeting credentials, thus granting itself more access to customer data that will eventually get exchanged as an encrypted key for ransom payments.

Selected Case Studies of Ransomware in the E-commerce Industry

Case Study One: Tupperware.com Ransomware Attack in 2021

Introduction to the Case

On 20 March 2020, multiple governmental organizations and private companies worldwide could not access the official website (tupperware.com) of Tupperware Brands Corporation, an American multinational direct-to-consumer company [7]. The issue captured more attention when the U.S. Federal Bureau of Investigation (FBI) suggested that the tupperware.com website was taken offline by a ransomware attack. Despite the purported ransomware incident, Tupperware Brands denied their website being hit by a ransomware attack. Eventually, a reputable ransomware gang named "REvil" published a story claiming that they encrypted tupperware.com files and asked the company to pay a ransom in exchange for a decryption tool [7].

CHAPTER 15   CASE STUDIES IN THE E-COMMERCE INDUSTRY

After discovering the ransomware, the organization disconnected the infected system from the network and later confirmed that sensitive data could not be stolen [7]. Tupperware issued a press release at the time, but the U.S. company did not inform the affected individuals. Tupperware did not notify affected persons of the incident of which it was actually or constructively aware in a reasonable period after discovering it. This practice was a common strategy to reduce the public relations harm such cybercrimes have on e-commerce companies in which customers have placed trust regarding their personal and financial data. Although the company did not divulge more information about what ransom figures were demanded, stakeholders were notified that hundreds of credit cards were compromised as the attackers confiscated their information.

Detailed Analysis

Since 2008, many e-commerce companies have relied on the Magento 1.x open-source e-commerce platform [7]. Unfortunately, this platform had lackluster cybersecurity features, which previous cybercriminal groups like the Magecart collective had exploited recently as 2016 [7]. The Tupperware.com ransomware attackers seemed to have fine-tuned their ability to undermine the e-commerce APIs, even selling these in ransomware-as-a-service (RaaS) trades within the cybercrime community [8]. However, on the day Tupperware.com's API was hacked, the attackers planted their malicious code in the Frequently Asked Questions (FAQs) portion of the API as a PNG file image [7]. Once users clicked on the interesting file image, the code activated and infected their application and devices, beginning to collect credit card information and other personal details. Notably, the malicious code triggered a fake payment form among users who had been hacked in the e-skimming criminal act, prompting them to enter their financial details about the credit card [9]. Eventually, the company's IT personnel noticed the attack because the attackers had only published the fake payment form in English.

Responses and Challenges to their Implementation

The first response by Tupperware.com once its IT personnel noticed the ransomware attack was to contact its API vendor. Tupperware management requested their vendor to conduct a malware scan to no avail, as it could not detect any. Tupperware had to spend valuable time repairing the malware infection to remove the ransomware threat. The vulnerability was detected on March 17, 2021, and by the end of the month, the website was up and running more securely than before [7]. The damaged parts had to be replaced, including installing a new e-commerce platform and financial information collection methods. Tupperware announced that no credit card details are usually stored on its e-commerce websites, and no social security number, ID, or IBAN were exposed.

The main challenge to implementing the recovery challenges intended to benefit Tupperware was its public denial of the ransomware attack. Although e-commerce companies regularly deny the existence of such attacks to protect their reputational integrity, such actions enable the attackers to continue harvesting victim data and gain further access to critical systems [10]. Some customers continue getting hacked because the company has not reported the presence of a cybersecurity attack. Second, the aging API has severe online security flaws, which was mentioned in previous industry reports and academic papers, but the company ignored these warnings.

Effectiveness of the Strategies

The strategies Tupperware.com used to respond to its ransomware attack were ineffective. First, denying an attack enables attackers to continue harvesting victim data and gaining access to critical systems [10]. Second, the waste of time in internal damage control could be remedied by outsourcing ransomware recovery to renowned and experienced

CHAPTER 15   CASE STUDIES IN THE E-COMMERCE INDUSTRY

cybersecurity companies [11]. Also, the company's choice of dishonesty in stating that the ransomware attackers confiscated no customer data affected its reputation in cybersecurity and the e-commerce industry.

Case Study Two: Ransomware Attack on Online Vendor X-Cart in October 2020
Introduction to the Case

Many online vendors in the growing e-commerce industry rely on third-party software for various objectives, from streamlining customer requests to inventory control and customer relationship management [13]. However, many of these third-party vendors need more software development budgets or could be more active in updating the security features of their products, exposing their clients to serious malware risks [14]. Such is the case in October 2020 when global vendor X-Cart could not gain access to or process store hosting systems [12]. A ransomware attack from one of these third-party vendors had infected its store hosting servers and systems, causing several stores to go offline. In contrast, others could not process customer email alerts–a critical customer service functionality among e-commerce vendors.

Upon further investigation, it was discovered that the ransomware attack was comparatively minimalistic, caused by substandard network security features in the third-party software from its rightful vendor [12]. Unfortunately, the outage lasted several days, forcing several store owners to consider class-action legal suits against the third-party software maker due to financial and reputational damages [12]. According to the online vendor, the attackers behind the ransomware attack did not provide further details. This is suspicious, considering these businesses are renowned for denying such attacks to avert public relations crises and customer loss of faith in cybersecurity.

CHAPTER 15   CASE STUDIES IN THE E-COMMERCE INDUSTRY

Detailed Analysis

X-Cart used third-party software from Seller Labs to optimize various functionalities in its e-commerce operations. These functionalities include streamlining customer inquiries and handling emails related to customer complaints, payment details, and product information [12]. However, the software was reported marred by several critical security gaps, which enabled even low-level cybercriminals to gain access as a training ground for their "hunt for bigger fish" [15]. Once the ransomware code infiltrated X-Cart's hosting servers through phishing or e-skimming, as demonstrated in Figure 15-2, it took control of the hosting servers' data and encrypted them. Due to the continuous access in online retail operations, the attackers had several instances of access to continue collecting additional information from the compromised servers [15]. Once hijacked, the servers were denied access to X-Cart store owners, management, and IT personnel for operational reasons.

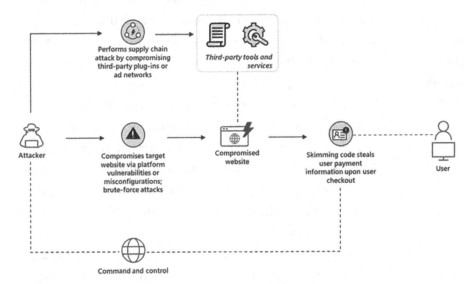

Figure 15-2. How E-skimming Works in Online Retail Environment [16]

Fortunately, the third-party software vendor had backed up the affected stores' data and began recovery actions immediately. Based on the execution observed and aversion to fiscal demands, one notices that maybe the X-Cart ransomware attack was training grounds because the execution of the attack resembles other e-commerce ransomware attacks [17]. However, the attackers failed to provide a method of contacting them or making ransom payments, signifying a lack of interest in fiscal benefits [12]. After several days of backup uploading and other recovery activities, X-Cart resumed operations in all its stores, with little information about the ransomware attacks on the media.

Responses and Challenges to Their Implementation

The first response by X-Cart employees upon realizing that their hosting servers were malfunctioning was informing their management. Subsequently, the company's parent company, Seller Labs, contacted their third-party software vendor, asking them to scan the product for malware [12]. The vendor confirmed and enacted industry-standard ransomware recovery procedures, including shutting off the entire software system [18]. Consequently, X-Cart could not handle various retail operations for several days while the vendor enacted recovery actions. Due to the absence of any contact information provided by the attackers, the vendor's IT team could not attempt negotiation or inquire about the attackers' demands [12]. The only course of action was to retrieve backups and begin uploading them into the improved and updated system. However, reputation and financial damage had already affected X-Cart in its competitive market space.

The main challenge in responding to this ransomware attack was a need for outside input from more skilled and experienced cybersecurity companies. Perhaps X-Cart and its software vendor handled the matter internally to prevent the media from getting word of the issue and increasing reputational damage to the firm [19]. Also, the attackers had not left contact information, undermining negotiation and recovery tactics involving communication with the crime's perpetrators [12]. Finally, retail operations could be more cohesive, with dozens and hundreds of stores making numerous server-host queries [20]. That means controlling operational downtime was difficult at the firm.

Effectiveness of the Strategies

The first response or recovery strategy was ineffective in assisting X-Cart to recover from its ransomware attack because engaging the services of more skilled and capable renowned cybersecurity companies during such attacks benefits the recovery with high-level skills [20]. Second, seeking the attackers' contact information failed due to lacking such information. The attackers were assumed to be less interested in fiscal gains and merely engaged in pen-testing for more significant attacks on larger organizations [21]. Finally, reverting to backup enabled the company to recover from the ransomware attack faster than comparative attacks, even with communication lines open between the attacker and victims [12]. The fact that X-Cart and its third-party software vendor had backups is a testament to proper ransomware recovery practices in the first place.

Case Study Three: VF Corporation's Ransomware Attack on December 13, 2023

Introduction to the Case

VF Corporation is a holding company whose apparel merchandise includes renowned brands like Timberland, Vans, and North Face. The company, based in Denver, Colorado, has both brick-and-mortar and online retail outlets, with a growing reliance on the latter due to lessons learned during the COVID-19 pandemics globally and the

CHAPTER 15 CASE STUDIES IN THE E-COMMERCE INDUSTRY

rapid digitalization of global retail in both fast fashion and the general apparel industries [23]. Notably, on December 13, 2023, the company's e-commerce IT team noted several unauthorized and unverified activities in its core digital systems responsible for various data-driven customer support and retail functionalities [22]. The team promptly reported these findings to the United States Securities and Exchange Commission (SEC) while promptly engaging in in-house and contracted cybersecurity management teams [22]. However, the ransomware attack resulted in the loss of data related to operational strategies, financial records, and personal identifying information about more than 35 million customers who had traded with the company [22]. Naturally, the attack vector remained unknown, according to executive managers at the e-commerce firm, who suspected they had concealed this information for reputational and market strategy reasons [24]. However, BLACK CAT claimed responsibility for the attack with no further details on ransom amounts or resolution process, if any [22].

Detailed Analysis

Industry analysis of the VF Corporation ransomware attack points toward a spear phishing or e-skimming attack to inject the malware code into the corporation's systems [25]. Once the ransomware code gained access, possibly at one of the online terminals on the customers' side or employees' internal systems, it immediately spread to all critical data resources with a customer-centric or financial implication [22]. IT managers from the company reported that the ransomware code encrypted those infected data systems, possibly making a ransom demand once the attack was discovered [26]. VF had robust malware protection in case of infection, and the IT team immediately enacted recovery protocols like closing their entire network and responding according to recovery frameworks after reporting to all significant statutory stakeholders, including the FBI [26].

CHAPTER 15 CASE STUDIES IN THE E-COMMERCE INDUSTRY

The attack targeted over 35 million customer data but failed to target any employee data or internal systems with reputational implications [22]. Therefore, experts suspected a dual attack where the ransomware perpetrators targeted the data alongside its systems and the company's brand by holding customer data hostage [27]. Therefore, if the ransom is unpaid, they release operational secrets to the cutthroat online retail market and sell sensitive customer data related to personal and financial information groups. However, the company's robust malware protections enabled it to upload existing backups, embolden vulnerable systems, and recover from all public relations damage while cooperating with law enforcement authorities [22].

Response and Challenges to Their Implementation

The first response from VF Corporation after the ransomware attack was to enact its internal ransomware recovery protocols, which included shutting down all online retail and sales operations [22]. Such strategies prevent the malware from spreading to sensitive portions of the company's network infrastructure. Concurrently, the company's IT team liaised with external cybersecurity experts in investigations and recovery optimization while backups were uploaded to restore network functionality, a widely embraced practice among victim organizations [27]. Also, the company's management reported the ransomware attack to the highest authorities, which is an effective method of forcing the perpetrators into a defensive position [22]. The last response seemed practical because the BLACK CAT cybercrime group claimed responsibility [22]. The company seemed to have paid because reports claimed its ransomware insurance coverage providers compensated it.

The main challenges related to implementing recovery strategies in this case study are aligned with online retail operational bottlenecks, with numerous access points and vast data generated [28]. Also, the servers handle a significant load regarding requests and customer data before, during, and after transactions. These challenges were partly managed by the decision to shut down the entire network to prevent malware spread and further network asset and data hijacks [26]. Nonetheless, the decision to report this ransomware to federal authorities has complicated negotiations as the insurance payout afterward indicated the ransomware attackers had considerable leverage.

Effectiveness of the Strategies

Apart from reporting its ransomware attack to the SEC, Department of Trade, and FBI, VF Corporation's strategies all seem adequate in complementing its recovery [22]. The decision to report may have instilled fear in the attackers, prompting them to make good their threats and forcing the company to pay. However, immediately shutting down all network functionalities prevented the e-commerce retailer from catastrophic network failure [26]. Also, engaging recovery experts from external specialists to cooperate with internal IT professionals enabled the company to recover using its ransomware recovery framework and protocols.

Case Study Four: Staples Ransomware Attack in December 2023
Introduction to the Case

In December 2023, office supplies, electronics, and furniture retailer Staples reported to law enforcement authorities that it suffered a ransomware attack [22]. Like other major retailers in the United States competitive online retail industry, Staples had been bolstering its e-commerce platforms to compete favorably with market giants like Amazon and eBay. Therefore, Staples executive management's reports shared intimated problems accessing and using order processing and delivery functionalities in its e-commerce operations network [22]. The ransomware attack also compromised Staples's customer service functionalities, affecting customer service operations and prompting all staff to be sent home as recovery efforts began [22]. Notably, the attack forced Zendesk, emails, and VPN-based employee portals to be shut down.

Detailed Analysis

According to reports from external experts focussed on the Staples ransomware attack, the most likely point of entry for the ransomware code was one of the employees' terminals through brute force attacks or phishing [22]. Immediately, the ransomware code infected the company's network; it targeted critical data and systems with design-based or operational vulnerabilities. Perhaps this is why the ransomware attack focussed on the order processing and delivery functionalities because of continuous access from employees, customers, and mutual connections with other critical databases like employee records and customer financial information [30]. These databases were encrypted as the company's IT team and external experts contracted to offer input scrambled to execute organizational and industry-standard ransomware recovery reactions [22]. Staples IT teams advise its personnel using the network to cease using Microsoft 365's functionality, which requires a single sign-off. This demonstrates the company's lack of prudence in using the application's full malware protection suite, as shown in Figure 15-3. This move may have protected the company from further damage as the ransomware only affected operations and did not infect customer data.

Response and Challenges Implementing Strategies

The first response by the Staples IT team was to shut down the company's entire e-commerce network to prevent further infection and the ransomware from affecting critical infrastructure [22]. Second, the IT professionals advised all workers with network access to cease using Microsoft 365's functionality, which requires a single sign-off [22]. Third, the company reported this cyberattack to law enforcement, affecting internal ransomware recovery protocols [22]. The only challenges observed in implementing the recovery strategies were the number of employees involved in the network shutdown [33]. The employees were supposed to understand the reason for not using their Microsoft 365 sign-in functionalities in an e-commerce company with operations all over the country.

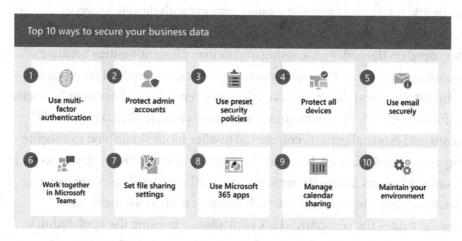

Figure 15-3. Microsoft 365 Malware Protections for Business Data [29]

Effectiveness of Strategies

Shutting down Staples immediately after the network and IT staff detected ransomware characteristics effectively stopped the spread of the code to critical infrastructure [31]. Also, the advice given to all employees with access to the company's e-commerce network and its Microsoft 365 single sign-in feature to abandon using it possibly prevented the attackers from launching advanced features of their ransomware code [27]. Among many e-commerce companies targeted recently in the United States, Staples was perhaps one of the most open about its ransomware attack [32]. Many online retailers hide technical and legal details of their attacks due to reputational and market risks. This decision enabled the company to benefit from pro-bono advice from industry experts and revealed the attack's vulnerabilities, making recovery more accessible and faster [32].

References

[1] S. Badotra and A. Sundas, "A systematic review on security of E-commerce systems," *International Journal of Applied Science and Engineering*, vol. 18, no. 2, pp. 1–19, 2021, doi: https://doi.org/10.6703/IJASE.202106_18(2).010.

[2] C. Metinko, "Cybersecurity Venture Funding Surpasses $20B In 2021, Fourth Quarter Smashes Record," *Crunchbase News*, Jan. 06, 2022. https://news.crunchbase.com/venture/cybersecurity-venture-funding-2021-record/

[3] H. F. Atlam and O. Oluwatimilehin, "Business Email Compromise Phishing Detection Based on Machine Learning: A Systematic Literature Review," *Electronics*, vol. 12, no. 1, p. 42, Dec. 2022, doi: https://doi.org/10.3390/electronics12010042.

[4] X. Liu *et al.*, "Cyber security threats: A never-ending challenge for e-commerce," *Frontiers in Psychology*, vol. 13, 2022, doi: https://doi.org/10.3389/fpsyg.2022.927398.

[5] I. Tariq, M. A. Sindhu, R. A. Abbasi, A. S. Khattak, O. Maqbool, and G. F. Siddiqui, "Resolving cross-site scripting attacks through genetic algorithm and reinforcement learning," *Expert Systems with Applications*, vol. 168, p. 114386, Apr. 2021, doi: https://doi.org/10.1016/j.eswa.2020.114386.

[6] A. Ali *et al.*, "Financial Fraud Detection Based on Machine Learning: A Systematic Literature Review," *Applied Sciences*, vol. 12, no. 19, p. 9637, 2022, doi: https://doi.org/10.3390/app12199637.

[7] N. Engineering, "The Latest e-Commerce Cyber Attacks and Their Implications for 2021," *Nox90*, Apr. 04, 2021. https://www.nox90.com/post/the-latest-e-commerce-cyber-attacks-and-their-implications-for-2021

[8] J. Howard, "Artificial intelligence: Implications for the Future of Work," *American Journal of Industrial Medicine*, vol. 62, no. 11, pp. 917–926, Aug. 2019, doi: https://doi.org/10.1002/ajim.23037.

[9] I. Castiglioni *et al.*, "AI applications to medical images: From machine learning to deep learning," *Physica Medica*, vol. 83, pp. 9–24, Mar. 2021, doi: https://doi.org/10.1016/j.ejmp.2021.02.006.

[10] M. M. Yamin, B. Katt, and V. Gkioulos, "Cyber ranges and security testbeds: Scenarios, functions, tools and architecture," *Computers & Security*, vol. 88, p. 101636, Jan. 2020, doi: https://doi.org/10.1016/j.cose.2019.101636.

[11] M. Humayun, N. Jhanjhi, A. Alsayat, and V. Ponnusamy, "Internet of things and ransomware: Evolution, mitigation and prevention," *Egyptian Informatics Journal*, vol. 22, no. 1, pp. 105–117, May 2020, doi: https://doi.org/10.1016/j.eij.2020.05.003.

[12] C. Cimpanu, "Ransomware hits e-commerce platform X-Cart," *ZDNET*, Nov. 08, 2020. https://www.zdnet.com/article/ransomware-hits-e-commerce-platform-x-cart/ (accessed Jun. 29, 2024).

[13] Y. K. Dwivedi *et al.*, "Artificial Intelligence (AI): Multidisciplinary Perspectives on Emerging challenges, opportunities, and Agenda for research, Practice and Policy," *International Journal of Information Management*, vol. 57, no. 101994, Aug. 2021, doi: https://doi.org/10.1016/j.ijinfomgt.2019.08.002.

[14] R. Venkateswaran, B. Ugalde, and T. Rogelio, "Impact of Social Media Application in Business Organizations," *International Journal of Computer Applications*, vol. 178, no. 30, pp. 5–10, Jul. 2019, doi: https://doi.org/10.5120/ijca2019919126.

[15] M. M. Ali and N. F. Mohd Zaharon, "Phishing—A Cyber Fraud: The Types, Implications and Governance," *International Journal of Educational Reform*, vol. 33, no. 1, p. 105678792210829, Mar. 2022, doi: https://doi.org/10.1177/10567879221082966.

[16] Microsoft Threat Intelligence, "Beneath the surface: Uncovering the shift in web skimming," *Microsoft Security Blog*, May 23, 2022. https://www.microsoft.com/en-us/security/blog/2022/05/23/beneath-the-surface-uncovering-the-shift-in-web-skimming/

[17] V. G. Dharmavaram, "Formjacking attack: Are we safe?," *Journal of Financial Crime*, vol. 28, no. 2, pp. 607–612, Oct. 2020, doi: https://doi.org/10.1108/jfc-07-2020-0138.

[18] S. Kamil, H. S. A. Siti Norul, A. Firdaus, and O. L. Usman, "The Rise of Ransomware: A Review of Attacks, Detection Techniques, and Future Challenges," *2022 International Conference on Business Analytics for Technology and Security (ICBATS)*, Feb. 2022, doi: https://doi.org/10.1109/icbats54253.2022.9759000.

[19] D. N. Molokomme, A. J. Onumanyi, and A. M. Abu-Mahfouz, "Edge Intelligence in Smart Grids: A Survey on Architectures, Offloading Models, Cyber Security Measures, and Challenges," *Journal of Sensor and Actuator Networks*, vol. 11, no. 3, p. 47, Aug. 2022, doi: https://doi.org/10.3390/jsan11030047.

[20] Rajeev Sobti, R. Garg, Ajeet Kumar Srivastava, and Gurpeet Singh Shahi, *Computer Science Engineering and Emerging Technologies*. CRC Press, 2024.

[21] M. Al-Hawawreh, M. Alazab, M. A. Ferrag, and M. S. Hossain, "Securing the Industrial Internet of Things against ransomware attacks: A comprehensive analysis of the emerging threat landscape and detection mechanisms," *Journal of Network and Computer Applications*, vol. 223, p. 103809, Mar. 2024, doi: https://doi.org/10.1016/j.jnca.2023.103809.

[22] Asimily.com, "4 Retail Cyberattacks that Hurt Businesses in 2023," *https://asimily.com*, Feb. 27, 2024. https://asimily.com/blog/4-retail-cyberattacks-that-hurt-businesses-2023/ (accessed Jun. 29, 2024).

[23] T. Dzhengiz, T. Haukkala, and O. Sahimaa, "(Un)Sustainable transitions towards fast and ultra-fast fashion," *Fashion and Textiles*, vol. 10, no. 1, May 2023, doi: https://doi.org/10.1186/s40691-023-00337-9.

[24] M. Lezzi, M. Lazoi, and A. Corallo, "Cybersecurity for Industry 4.0 in the current literature: A reference framework," *Computers in Industry*, vol. 103, pp. 97–110, Dec. 2018, doi: https://doi.org/10.1016/j.compind.2018.09.004.

[25] S. Chowdhury, S. Mukherjee, Saranya Naha Roy, R. Mehdi, and R. Banerjee, "An overview of cybersecurity risks during the COVID-19 pandemic period," *Scientific Voyage*, vol. 1, no. 3, pp. 47–54, Sep. 2020.

[26] T. McIntosh, A. S. M. Kayes, Y.-P. P. Chen, A. Ng, and P. Watters, "Ransomware Mitigation in the Modern Era: A Comprehensive Review, Research Challenges, and Future Directions," *ACM Computing Surveys*, vol. 54, no. 9, pp. 1–36, Dec. 2022, doi: https://doi.org/10.1145/3479393.

[27] Ö. A. Aslan and R. Samet, "A Comprehensive Review on Malware Detection Approaches," *IEEE Access*, vol. 8, pp. 6249–6271, Jan. 2020, doi: https://doi.org/10.1109/ACCESS.2019.2963724.

[28] C. Beaman, A. Barkworth, T. D. Akande, S. Hakak, and M. K. Khan, "Ransomware: Recent advances, analysis, challenges and future research directions," *Computers & Security*, vol. 111, no. 1, Dec. 2021, doi: https://doi.org/10.1016/j.cose.2021.102490.

[29] denisebmsft, "Microsoft 365 for business security best practices - Microsoft 365 Business Premium," *learn.microsoft.com*, Jul. 11, 2023. https://learn.microsoft.com/en-us/microsoft-365/business-premium/secure-your-business-data?view=o365-worldwide

[30] F. Aldauiji, O. Batarfi, and M. Bayousif, "Utilizing Cyber Threat Hunting Techniques to Find Ransomware Attacks: A Survey of the State of the Art," *IEEE Access*, no. 13, pp. 1–1, 2022, doi: https://doi.org/10.1109/access.2022.3181278.

[31] S. R. Matthijsse, M. S. van 't Hoff-de Goede, and E. R. Leukfeldt, "Your files have been encrypted: a crime script analysis of ransomware attacks," *Trends in Organized Crime*, Apr. 2023, doi: https://doi.org/10.1007/s12117-023-09496-z.

[32] M. Gazzan and F. T. Sheldon, "Opportunities for Early Detection and Prediction of Ransomware Attacks against Industrial Control Systems," *Future Internet*, vol. 15, no. 4, p. 144, Apr. 2023, doi: https://doi.org/10.3390/fi15040144.

[33] Sangfor Technologies, "A Comprehensive List of Top Ransomware Attacks in 2023," *Sangfor Technologies*, Dec. 21, 2023. https://www.sangfor.com/blog/cybersecurity/list-of-top-ransomware-attacks-in-2023

CHAPTER 16

Ransomware, Inc.: The Business and Economics of Digital Extortion

Author:
Varun Garde

Introduction: The Ransomware Economy

In October 2023, the British Library fell victim to a crippling ransomware attack [1]. This incident, which disrupted services, compromised data, and forced the library to undertake a costly recovery process, exemplifies the growing threat posed by the ransomware economy.

Ransomware, a type of malware that encrypts files and demands payment for decryption, has become a highly profitable business model for cybercriminals. In 2021, global losses from ransomware were estimated at $20 billion, with the average ransom payment exceeding $500,000 [2].

CHAPTER 16 RANSOMWARE, INC.: THE BUSINESS AND ECONOMICS OF DIGITAL EXTORTION

The true cost of ransomware, however, extends beyond the ransom itself, encompassing lost productivity, reputational damage, and the expense of rebuilding compromised systems.

As the ransomware threat evolves, it is crucial for organizations to understand the economic drivers and ecosystem behind these attacks. This chapter explores the key players, business models, and growth factors of the ransomware economy while identifying potential strategies for negating the advantages of attackers.

Inside the Ransomware Marketplace

From Petty Crime to Big Business: The Evolution of Ransomware

Ransomware has undergone a dramatic transformation from its early days of small-scale, opportunistic attacks to the highly organized, lucrative, and professionalized criminal enterprise it is today. This evolution has been fueled by several key developments:

1. The rise of cryptocurrency, providing an anonymous and difficult-to-trace means of payment
2. The increasing availability of powerful, easy-to-use ransomware tools on the dark web
3. The shift toward targeted attacks on organizations more likely to pay larger ransoms, such as businesses, healthcare providers, and government agencies
4. The growing sophistication of encryption techniques used by ransomware, making it more difficult for victims to recover their data without paying the ransom.

CHAPTER 16   RANSOMWARE, INC.: THE BUSINESS AND ECONOMICS OF DIGITAL EXTORTION

As the ransomware market matured, cybercriminal groups began to differentiate themselves through branding, offering customer support, and issuing corporate-style communications [4]. The emergence of Ransomware-as-a-Service (RaaS) has further accelerated this professionalization by lowering barriers to entry for attackers and fostering a thriving ecosystem of developers, affiliates, and service providers.

The Ransomware Value Chain: Players, Roles, and Profit Shares

One of the most striking features of the modern ransomware economy is the degree of specialization and collaboration among its participants. Rather than operating in isolation, ransomware actors typically work within a complex, interdependent value chain that includes a range of players with distinct roles and expertise.

At the core of this value chain are the ransomware developers, who create the malware strains used in attacks. These developers may work independently or as part of larger cybercriminal organizations, and they often license their creations to other groups through RaaS arrangements.

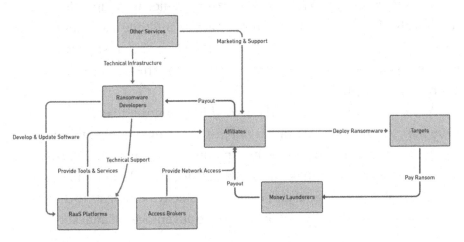

Figure 16-1. *The Ransomware Value Chain*

331

CHAPTER 16   RANSOMWARE, INC.: THE BUSINESS AND ECONOMICS OF DIGITAL EXTORTION

RaaS has become an increasingly popular business model in the ransomware ecosystem, as it allows developers to monetize their skills without having to carry out attacks themselves. Under a typical RaaS arrangement, the developer provides the malware and infrastructure, while affiliates, who are essentially the "customers" of the service, are responsible for distributing the malware and carrying out the actual attacks [5].

The profits from successful attacks are then split between the developer and the affiliate, with the developer typically receiving a smaller share (e.g., 20-30%) in exchange for providing the tools and support. This model has proven to be highly effective, as it allows both developers and affiliates to specialize and benefit from economies of scale.

In addition to developers and affiliates, the ransomware value chain also includes a range of supporting players, such as access brokers, who sell compromised network access to attackers; money launderers, who help to conceal and clean the proceeds of ransomware campaigns; and even marketers and public relations specialists, who help to promote RaaS offerings and manage the reputations of cybercriminal brands [6].

This elaborate division of labor and profit sharing has enabled the ransomware economy to become increasingly efficient, adaptable, and resilient. By distributing risk and reward across multiple actors and allowing each to focus on their core competencies, the ransomware ecosystem has been able to thrive and innovate in the face of growing efforts to combat it.

As we explore the inner workings of this underground economy, it becomes clear that ransomware is not just a technical challenge, but a complex, multifaceted business that operates according to its own distinct logic and incentives. Understanding this business model and the roles of its various players is crucial for developing effective strategies to disrupt the ransomware value chain and negate the advantages of attackers.

CHAPTER 16 RANSOMWARE, INC.: THE BUSINESS AND ECONOMICS OF DIGITAL EXTORTION

Tailoring the Ransom Note: The Art and Science of Pricing Extortion

Setting the right ransom price is crucial for attackers, as they aim to maximize profits while minimizing the risk of non-payment or detection. Ransomware groups often conduct extensive research on their targets before launching an attack, analyzing factors such as the victim's financial resources, the value of the encrypted data, and market dynamics.

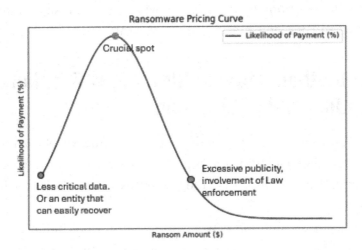

Figure 16-2. Few Decisions Are as Crucial as Setting the Ransom Price

Some groups even employ machine learning algorithms to predict the optimal ransom amount for each victim [7]. The result is a highly tailored ransom demand designed to exert maximum pressure while avoiding the impression of being unreasonable.

Ransomware groups must also consider the long-term reputational risks associated with their pricing strategies. Demanding too much can alienate victims and attract unwanted attention from law enforcement, while asking for too little may undermine their credibility and bargaining power.

To navigate these competing pressures, some groups have begun experimenting with more sophisticated pricing strategies, such as dynamic pricing algorithms that adjust the ransom in real-time based on the victim's behavior and market conditions [9]. Others have diversified their revenue streams by offering additional services, such as data exfiltration and threats of public leaks, to extract higher payments.

As the ransomware market evolves, pricing strategies are likely to become even more complex and data driven. Understanding the economic calculus behind ransom demands can provide valuable insights into the motivations and decision-making processes of attackers.

Attack Methodology: Infiltration, Exfiltration, Encryption, and Destruction

While ransomware tactics constantly evolve, most attacks follow a similar methodology. Understanding the key stages of a ransomware attack can help organizations better prepare their defenses and minimize the impact of a breach.

The first stage is infiltration, where the attacker gains initial access to the target network through phishing emails, compromised credentials, or unpatched vulnerabilities [10]. Once inside, the attacker seeks to elevate their privileges and move laterally to identify and compromise critical assets and data.

In many cases, the attacker will attempt to exfiltrate sensitive data before deploying the ransomware payload, a tactic known as "double extortion" [11]. This involves threatening to release the stolen data publicly if the ransom is not paid, adding further pressure on the victim.

Once the attacker has compromised the desired targets, they deploy the ransomware, encrypting the victim's files and displaying a ransom note with payment instructions. The encryption process is designed to be fast and efficient, using strong cryptographic algorithms that are virtually impossible to break without the attacker's private key.

CHAPTER 16   RANSOMWARE, INC.: THE BUSINESS AND ECONOMICS OF DIGITAL EXTORTION

In some cases, the attacker may also attempt to destroy or disrupt the victim's backup systems and recovery mechanisms, making it more difficult to restore data without paying the ransom. This tactic was used in the British Library attack, causing significant damage to its IT infrastructure [1].

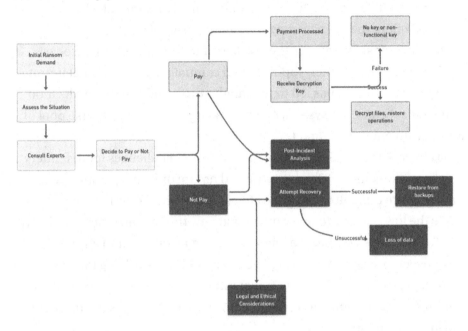

Figure 16-3. *Key Considerations and Tradeoffs Involved in Deciding Whether to Pay a Ransom, and the Potential Consequences*

The destructive nature of modern ransomware attacks highlights the importance of a comprehensive incident response plan. Organizations that can quickly detect, contain, and recover from an attack are more likely to minimize the impact on their operations.

However, as the British Library case demonstrates, even well-prepared organizations can struggle to recover from a sophisticated attack, underscoring the need for a holistic approach to ransomware defense that encompasses technical controls, processes, and governance.

CHAPTER 16 RANSOMWARE, INC.: THE BUSINESS AND ECONOMICS OF DIGITAL EXTORTION

The Emergence of Ransomware as a Service

Ransomware as a Service (RaaS) has been a significant development in the ransomware economy, expanding the accessibility and profitability of attacks. This business model allows developers to lease or sell their malware to affiliates, who then distribute it and carry out attacks.

RaaS offers several advantages for both parties. Developers can monetize their skills without the risk of conducting attacks themselves, while affiliates can launch sophisticated campaigns with minimal upfront investment [12]. This has led to the rapid growth and professionalization of the ransomware ecosystem, enabling a larger and more diverse pool of attackers to enter the market.

However, the RaaS model also presents new challenges for defenders and law enforcement. The decentralized nature of RaaS operations can make it more difficult to attribute attacks and disrupt criminal networks, while the low barriers to entry mean that new threats can emerge quickly.

To effectively combat the RaaS threat, organizations and authorities must develop strategies that can keep pace with the evolving ransomware landscape, such as investing in threat intelligence, collaborating to disrupt criminal infrastructure, and creating new legal frameworks to prosecute ransomware actors.

Ransomware Revenue and Profitability

The true scale of the ransomware economy is difficult to measure due to its secretive and illicit nature. However, estimates suggest that annual global losses are in the billions of dollars and growing rapidly. In 2021, Cybersecurity Ventures predicted that ransomware damages would reach $20 billion, up from $11.5 billion in 2019 [2].

At the individual level, ransomware can be highly profitable. In 2020, the average ransom payment was estimated at over $300,000 [14]. Successful RaaS developers and affiliates can earn millions of dollars per year.

However, the economics of ransomware are complex. Not all victims pay, and those who do may not recover their data fully. Ransomware groups also face significant operational costs and risks, such as maintaining infrastructure and evading law enforcement.

As the market becomes more crowded, some groups have struggled to maintain profitability. In response, many have turned to more aggressive tactics, such as targeting critical infrastructure and using multiple extortion methods.

These trends suggest that the ransomware economy is maturing and consolidating, with the most successful groups seeking to expand their market share. For organizations defending against these threats, understanding the financial drivers and incentives of the ransomware market is essential for developing effective countermeasures and disruption strategies.

The Minds Behind the Malware: Incentives of Ransomware Actors

The Profit Motives and Risk-Return Tradeoffs of Ransomware Actors

Ransomware actors at every level of the value chain seek to maximize profits while minimizing exposure to legal, financial, and operational risks. For developers and affiliates, this means carefully balancing the potential payoffs of an attack against the costs and risks of detection, disruption, and prosecution.

Targeting high-value victims, such as those in critical sectors like healthcare and finance, can yield substantial payouts but also comes with significant risks. These organizations often have robust cybersecurity defenses and are more likely to engage with law enforcement.

CHAPTER 16   RANSOMWARE, INC.: THE BUSINESS AND ECONOMICS OF DIGITAL EXTORTION

To mitigate these risks, ransomware actors must constantly adapt their tactics, investing in new tools and infrastructure to stay ahead of defenders. These investments can be costly and time-consuming, cutting into potential profits.

Ransomware actors must also contend with the risk of disruption and prosecution by law enforcement agencies. As the impact of ransomware attacks has grown, so too has the response from governments and international organizations, with many countries treating ransomware as a national security threat.

To avoid detection and arrest, ransomware actors employ various measures to conceal their identities and activities, such as using anonymous communication channels and laundering illicit proceeds through complex networks of shell companies and cryptocurrencies. These measures can be expensive and cumbersome, adding to the overall cost and complexity of conducting a ransomware campaign.

Despite these risks and challenges, the potential profits from ransomware remain a powerful incentive for many actors. As long as these incentives remain in place, the ransomware threat will likely continue to grow and evolve.

The Role of Cryptocurrency in the Ransomware Ecosystem

Cryptocurrencies have been a key enabler of the modern ransomware economy, providing attackers with a means for receiving and laundering ransom payments while making it harder for authorities to track and recover funds.

Cryptocurrencies offer several advantages for ransomware actors, including the ability to receive payments from victims worldwide without relying on traditional banking channels. They also provide a degree of anonymity and privacy, making it harder to trace funds and identify individuals behind a campaign.

In addition to enabling payments, cryptocurrencies play a key role in the laundering process for ransomware profits. After receiving a payment, attackers typically use techniques like mixing services, tumblers, and chain hopping to conceal the origin and destination of the funds before converting them into fiat currency or other assets.

However, the use of cryptocurrencies also comes with risks for attackers. As the impact of ransomware has grown, governments and law enforcement have begun to focus more on the role of cryptocurrencies in facilitating these crimes, leading to increased scrutiny of exchanges and the development of new tracing and recovery tools.

In some cases, authorities have successfully seized cryptocurrency payments from ransomware attacks by exploiting vulnerabilities in the attackers' laundering processes. To mitigate these risks, some groups have begun experimenting with alternative payment methods, such as privacy-enhanced cryptocurrencies or requiring payments in non-traditional assets.

Deterring Ransomware: The Limits of Economic Incentives

While economic incentives are a key factor in the ransomware ecosystem, a purely economic approach to deterring ransomware may be insufficient or even counterproductive.

First, not all individuals involved in ransomware are motivated solely by financial gain. Some actors may be driven by ideological or political goals, the desire for notoriety, or intellectual curiosity. For these individuals, strategies that focus solely on altering financial incentives may have limited impact.

Second, even for financially motivated actors that are desperate enough, the potential rewards of a successful attack may outweigh significant risks and costs. With some groups earning tens of millions of dollars per year, the lure of profits may be strong enough for some to try. And where many try, some may succeed.

CHAPTER 16   RANSOMWARE, INC.: THE BUSINESS AND ECONOMICS OF DIGITAL EXTORTION

Additionally, the decentralized and globalized nature of the ransomware ecosystem makes it difficult to impose meaningful costs and risks on all actors. While some jurisdictions may have strong laws and enforcement capabilities, others may have weaker or more permissive regimes, creating safe havens for attackers.

There is also a risk that overly aggressive attempts to deter ransomware through economic means could backfire. Blanket bans on cryptocurrency transactions have been considered and all but put on hold because of its impact on legitimate activity. Strict policies against ransom payments could also lead to more destructive and coercive tactics.

As such, while economic incentives are important, they must be balanced against other considerations. A comprehensive strategy to combat ransomware should combine economic measures with other approaches, such as improving defenses, strengthening international cooperation, and investing in education and awareness.

Ransomware and the Bottom Line: The Business Costs of Digital Extortion

Ransomware attacks are not just a technical challenge for organizations–they are also a major business and financial threat. The direct and indirect costs of a successful ransomware attack can be staggering, ranging from the immediate loss of productivity and revenue to the long-term damage to an organization's reputation and customer trust.

The Direct Financial Costs of Ransomware

The most visible cost of a ransomware attack is often the ransom payment itself. However, even for organizations that choose to pay, the ransom is just the tip of the iceberg in terms of financial impact.

CHAPTER 16   RANSOMWARE, INC.: THE BUSINESS AND ECONOMICS OF DIGITAL EXTORTION

In addition to the ransom, organizations face a range of direct costs, including

1. Incident response and investigation
2. System recovery and restoration
3. Legal and regulatory expenses
4. Increased cybersecurity investments

The British Library's experience illustrates the scale of these costs. In the wake of the attack, the library had to hire external experts, invest in new hardware and software to rebuild its infrastructure, and address potential legal and regulatory issues related to data loss [1].

The Indirect Costs and Long-Term Impacts of Ransomware

Beyond the direct financial costs, ransomware attacks can have significant indirect and long-term impacts on an organization's operations, reputation, and competitiveness. These costs, while harder to quantify, are often just as significant as the direct costs.

Some key indirect costs and impacts include

1. Operational disruptions and lost productivity
2. Reputational damage and loss of customer trust
3. Competitive disadvantage and market share loss

In the case of the British Library, the attack disrupted access to services and resources for months, affecting staff and patrons and leading to a loss of trust among some stakeholders [1].

CHAPTER 16   RANSOMWARE, INC.: THE BUSINESS AND ECONOMICS OF DIGITAL EXTORTION

The Business Case for Investing in Ransomware Prevention and Mitigation

Given the significant financial and operational risks posed by ransomware, there is a strong business case for organizations to invest in prevention and mitigation measures. While these investments can be costly in the short term, they can pay significant dividends in the long term by reducing the likelihood and impact of successful attacks.

Some key areas for investment include

- **Cybersecurity Technologies and Services**: Investing in robust cybersecurity technologies and services, such as endpoint protection, network monitoring, and threat intelligence, can help organizations detect and prevent ransomware attacks before they cause significant damage. Working with experienced cybersecurity providers can also provide access to specialized expertise and resources that may be difficult to develop in-house.

- **Employee Education and Awareness**: Investing in employee education and awareness programs can help reduce the risk of human error and social engineering, which are common vectors for ransomware attacks. Training employees to recognize and report suspicious emails, attachments, and links can be a cost-effective way to improve an organization's overall cybersecurity posture.

- **Incident Response and Disaster Recovery Planning**: Developing and testing comprehensive incident response and disaster recovery plans can help organizations minimize the impact of successful ransomware attacks and recover more quickly from disruptions. This can include regular backups, offline data storage, and failover systems to ensure continuity of critical operations.

- **Cybersecurity Insurance**: Investing in cybersecurity insurance can provide financial protection against the costs of ransomware attacks, including ransom payments, incident response, and legal and regulatory expenses. However, it is important for organizations to carefully review the terms and conditions of their policies to ensure adequate coverage and to understand any exclusions or limitations.

Ultimately, the key to building a strong business case for ransomware prevention and mitigation is to understand the unique risks and impacts facing your organization and to develop a comprehensive strategy that balances the costs of investment against the potential costs of inaction. By taking a proactive and risk-based approach to cybersecurity, organizations can reduce their exposure to ransomware and other cyberthreats, while also building resilience and agility in the face of an ever-evolving threat landscape.

Hacking the Hackers: Economic Strategies to Combat Ransomware

Given the significant economic incentives that drive the ransomware ecosystem, there is growing recognition among cybersecurity experts and policymakers that combating this threat will require more than just technical solutions. To truly disrupt the ransomware business model and deter future attacks, we must also develop economic strategies that raise the costs and risks for attackers, while reducing the potential rewards.

CHAPTER 16   RANSOMWARE, INC.: THE BUSINESS AND ECONOMICS OF DIGITAL EXTORTION

Reducing the Profitability of Ransomware

One key approach to combating ransomware is to reduce the profitability of attacks, making it less financially attractive for cybercriminals to engage in this activity. There are several ways this could be achieved:

1. **Encouraging Organizations not to Pay Ransoms**: This is easier said than done, particularly for organizations that rely on access to critical data and systems. But a rational analysis shows that paying ransom only signals the payout probability and overall vulnerability of the victim, encouraging future attacks and funding for cybercriminals to develop more sophisticated tools and techniques. By adopting a collective stance against ransom payments, organizations can help to reduce the overall profitability of the ransomware ecosystem.

2. **Disrupting Cryptocurrency Transactions**: As noted earlier, cryptocurrencies play a key role in enabling ransomware transactions and facilitating the laundering of illicit proceeds. By working with cryptocurrency exchanges, financial institutions, and law enforcement agencies to identify and disrupt suspicious transactions, it may be possible to make it harder for cybercriminals to receive and use ransom payments. This could include measures such as increased know-your-customer (KYC) requirements for cryptocurrency exchanges, enhanced monitoring and reporting of suspicious transactions, and the development of new tools and techniques for tracing and seizing illicit funds.

3. **Targeting the Infrastructure and Services That Support Ransomware**: Another way to reduce the profitability of ransomware is to target the infrastructure and services that support the ransomware ecosystem, such as bulletproof hosting providers, dark web marketplaces, and money laundering services. By disrupting these critical enablers, it may be possible to make it harder for cybercriminals to operate and scale their ransomware campaigns. This could include measures such as increased law enforcement action against cybercrime infrastructure, enhanced international cooperation and information sharing, and the development of new legal and regulatory frameworks to address the unique challenges posed by the ransomware ecosystem.

Increasing the Costs and Risks for Ransomware Actors

In addition to reducing the profitability of ransomware, another key strategy is to increase the costs and risks for the individuals and groups involved in this activity. This could include:

1. **Strengthening Law Enforcement and Prosecution**: This means increased funding and resources for cybercrime investigations, enhanced training and capacity building for law enforcement agencies, and the development of new legal and regulatory frameworks to address the unique challenges posed by ransomware and other forms of cybercrime. By increasing the likelihood of detection, arrest,

and prosecution, it may be possible to deter some individuals from engaging in ransomware activity and to disrupt the operations of established ransomware groups.

2. **Imposing Financial Sanctions and Asset Seizures**: This could include measures such as freezing bank accounts and cryptocurrency wallets, seizing physical assets such as property and vehicles, and imposing travel bans and other restrictions on key ransomware actors. By making it harder for cybercriminals to access and use their illicit proceeds, it may be possible to disrupt the financial incentives that drive the ransomware ecosystem.

3. **Conducting Offensive Cyber Operations**: This could include measures such as hacking and infiltrating ransomware groups, sabotaging their malware and infrastructure, and exposing their operations and identities to law enforcement and the public. Offensive cyber operations can be controversial and carry significant risks–perhaps more risk than rewards in regions with weaker checks and balances, they may signal the high cost of illegal activity when conditions are right.

Building Resilience and Reducing the Impact of Ransomware

Finally, in addition to reducing profitability and increasing the costs and risks of ransomware, it is also important to build resilience and reduce the impact of successful attacks. This could include:

CHAPTER 16   RANSOMWARE, INC.: THE BUSINESS AND ECONOMICS OF DIGITAL EXTORTION

1. **Improving Cybersecurity Defenses and Incident Response**: One of the most effective ways to reduce the impact of ransomware is to improve organizations' cybersecurity defenses and incident response capabilities. This could include measures such as implementing strong access controls and network segmentation, conducting regular vulnerability assessments and penetration testing, and developing and testing incident response and disaster recovery plans. By making it harder for ransomware to spread and cause damage, and by enabling faster and more effective response and recovery, organizations can reduce the overall impact of successful attacks.

2. **Promoting Information Sharing and Collaboration**: Another key strategy for building resilience against ransomware is to promote information sharing and collaboration among organizations, cybersecurity providers, and government agencies. This could include measures such as establishing threat intelligence sharing platforms, developing industry-specific best practices and standards, and fostering public-private partnerships to address shared cybersecurity challenges. By working together to share information and resources, organizations can improve their collective defense against ransomware and other cyberthreats.

3. **Investing in Research and Development:** Finally, it is important to invest in ongoing research and development to stay ahead of the evolving ransomware threat. This could include measures such as developing new cybersecurity technologies and techniques, studying the tactics and motivations of ransomware actors, and exploring new economic and policy approaches to combat this threat. By continually innovating and adapting to the changing threat landscape, we can help to build a more secure and resilient digital ecosystem for the future.

Case Study: The Colonial Pipeline Attack and the US Government's Response

One of the most high-profile ransomware attacks in recent years was the May 2021 attack on Colonial Pipeline, which disrupted fuel supplies across the southeastern United States and prompted a strong response from the US government. This case study illustrates some of the economic strategies and challenges involved in combating ransomware at a national level.

The attack on Colonial Pipeline was carried out by a ransomware group known as DarkSide, which used a sophisticated strain of ransomware to encrypt the company's computer systems and demand a ransom payment of approximately $4.4 million in Bitcoin [16]. The attack caused Colonial Pipeline to shut down its operations for several days, leading to fuel shortages and price spikes across the affected region.

In response to the attack, the US government took several significant steps to disrupt the ransomware ecosystem and deter future attacks. These included

CHAPTER 16 RANSOMWARE, INC.: THE BUSINESS AND ECONOMICS OF DIGITAL EXTORTION

1. **Seizing the Ransom Payment**: In June 2021, the US Department of Justice announced that it had seized approximately $2.3 million in Bitcoin that had been paid by Colonial Pipeline as a ransom to the DarkSide group [15]. This seizure was made possible by the FBI's infiltration of DarkSide's cryptocurrency wallet and the use of blockchain analysis tools to trace the ransom payment. By demonstrating the ability to seize ransom payments even after they have been made, the US government sent a powerful message to ransomware groups that their profits are not safe from law enforcement.

2. **Imposing Sanctions on Ransomware Groups**: In September 2021, the US Treasury Department imposed sanctions on the DarkSide ransomware group and several associated individuals and entities, including a Russian cryptocurrency exchange that had facilitated ransomware transactions [17]. These sanctions made it illegal for US persons to engage in transactions with the designated entities and individuals, and also signaled the US government's willingness to use financial measures to disrupt the ransomware ecosystem.

3. **Launching a Multiagency Ransomware Task Force**: In July 2021, the US government launched a multiagency ransomware task force, led by the Department of Justice, to coordinate the federal response to the growing ransomware threat [18]. The task force was charged with developing strategies to disrupt ransomware groups, improve

CHAPTER 16   RANSOMWARE, INC.: THE BUSINESS AND ECONOMICS OF DIGITAL EXTORTION

incident response and resilience, and promote international cooperation and information sharing. By bringing together expertise and resources from across the government, the task force aimed to create a more comprehensive and effective approach to combating ransomware.

While the US government's response generally seen as setting a new standard, it also highlighted some of the challenges and limitations of these approaches. For example:

1. **The Limits of Ransom Seizures**: While the seizure of the Colonial Pipeline ransom payment was a significant achievement, it also highlighted the difficulties of recovering ransom payments once they have been made. In this case, the FBI was able to seize the funds because they had already infiltrated DarkSide's cryptocurrency wallet, but in many other cases, ransom payments may be much harder to trace and recover, particularly if they are made using privacy-enhanced cryptocurrencies or other anonymizing techniques.

2. **The Challenges of International Cooperation**: While the US government's sanctions against DarkSide and associated entities sent a strong message, they also highlighted the challenges of combating ransomware groups that operate across international borders. Many ransomware groups are based in countries with weak law enforcement and limited cooperation with the United States, making it difficult to pursue criminal charges or seize

assets. Improving international cooperation and developing new legal and regulatory frameworks to address these challenges will be critical for effective ransomware prevention and response.

3. **The Need for Private Sector Engagement**: While the US government played a key role in responding to the Colonial Pipeline attack, it also highlighted the importance of private sector engagement in combating ransomware. Colonial Pipeline ultimately decided to pay the ransom demand on its own, without government involvement, and many other companies may be reluctant to involve law enforcement in ransomware incidents due to concerns about reputational damage, legal liability, or operational disruption. Building trust and cooperation between the private sector and government will be essential for improving ransomware incident response and resilience.

Overall, the Colonial Pipeline attack and the US government's response demonstrate both the potential and the challenges of using economic strategies to combat ransomware. While measures such as ransom seizures, sanctions, and multi-agency coordination can help to disrupt the ransomware ecosystem and deter future attacks, they must be part of a comprehensive and adaptable approach that also includes improving cybersecurity defenses, building international cooperation, and engaging the private sector as a key partner in the fight against ransomware.

Weathering the Storm: Lessons in Organizational Resilience

As the ransomware threat continues to evolve and expand, organizations must not only invest in prevention and mitigation measures but also cultivate a culture of resilience and adaptability. This means being prepared to weather the storm of a successful attack, to learn from the experience, and to emerge stronger and more secure on the other side.

Developing a Comprehensive Incident Response Plan

A critical element of resilience is having a comprehensive and well-tested incident response plan. This plan should outline specific steps and procedures to be followed in the event of an attack, including roles and responsibilities, detection and analysis, containment and eradication, recovery and restoration, and post-incident review and improvement.

To be effective, the plan must be regularly tested and updated to ensure it remains relevant in the face of evolving threats and changing organizational needs.

Building a Culture of Security Awareness and Responsibility

Organizations must also cultivate a culture of security awareness and responsibility among all employees and stakeholders. This involves providing regular training and education, encouraging openness and transparency around security issues, and empowering employees to be part of the solution.

CHAPTER 16   RANSOMWARE, INC.: THE BUSINESS AND ECONOMICS OF DIGITAL EXTORTION

By building a culture of awareness and responsibility, organizations can create a more resilient workforce better equipped to prevent, detect, and respond to ransomware and other threats.

Embracing Continuous Learning and Improvement

To truly cultivate resilience, organizations must embrace a mindset of continuous learning and improvement. This means conducting regular risk assessments and audits, investing in research and development, sharing lessons learned and best practices, and being willing to adapt based on new threats and technologies.

The case of the British Library's response to its October 2023 ransomware attack provides a useful example of organizational resilience in action. Despite suffering significant disruption and damage to its IT systems and services, the library was able to weather the storm and emerge stronger and more secure. Key elements of its response included [1]:

- Activating its existing crisis management and major incident response plans, which provided a clear framework and structure for decision making and communication during the incident.

- Engaging external expertise and support from the National Cyber Security Centre (NCSC) and other partners to investigate the incident and develop a recovery plan.

- Communicating openly and transparently with staff, users, and stakeholders about the impact of the incident and the steps being taken to restore services, while also managing expectations about the timeline for recovery.

CHAPTER 16   RANSOMWARE, INC.: THE BUSINESS AND ECONOMICS OF DIGITAL EXTORTION

- Conducting a comprehensive post-incident review to identify lessons learned and areas for improvement and using the findings to inform a longer-term "Rebuild & Renew" program to create a more resilient and secure IT infrastructure and culture.

By embracing a culture of resilience, learning, and adaptation, the British Library was able to not only recover from the ransomware attack but also to seize the opportunity to transform and strengthen its operations and services for the future.

Conclusion: Navigating the Future of Ransomware

As we have seen throughout this chapter, the ransomware threat is a complex and evolving challenge that requires a multifaceted and adaptive response. From understanding the economic drivers and ecosystem behind ransomware attacks to developing effective prevention and mitigation strategies, there is no simple solution or silver bullet.

However, by adopting a proactive and collaborative approach that combines technical, economic, and organizational measures, we can begin to turn the tide against ransomware and build a more secure and resilient digital future. This will require ongoing effort and investment from all stakeholders, including governments, businesses, academia, and civil society.

Some key priorities for the future of ransomware defense and response include

1. **Improving International Cooperation and Coordination**: Ransomware is a global threat that requires a global response. Governments and international organizations must work together to

harmonize legal and regulatory frameworks, share intelligence and best practices, and coordinate joint operations against ransomware groups and infrastructure.

2. **Strengthening Public-Private Partnerships**: Combating ransomware will require close collaboration between the public and private sectors, including sharing threat intelligence, developing joint response plans, and investing in research and development. Governments can play a key role in facilitating these partnerships and providing incentives for private sector participation.

3. **Investing in Education and Awareness**: Improving cybersecurity education and awareness at all levels, from individual users to corporate executives and policymakers, will be critical for building a more cyberliterate and resilient society. This should include both technical training and broader education on the social, economic, and political dimensions of cybersecurity.

4. **Developing New Technologies and Techniques**: As ransomware groups continue to innovate and adapt, so too must defenders. This will require ongoing investment in research and development of new technologies and techniques for prevention, detection, and response, including advanced encryption, machine learning, and blockchain-based solutions.

5. **Cultivating a Culture of Resilience and Adaptation:** Finally, organizations and individuals must cultivate a culture of resilience and adaptation in the face of the ransomware threat. This means being prepared for the inevitable attacks that will occur, having the plans and capabilities in place to respond effectively, and being willing to learn and adapt based on experience.

The fight against ransomware is far from over, and the road ahead will undoubtedly be challenging. But by working together and embracing a spirit of resilience, innovation, and collaboration, we can build a future in which the benefits of digital technology are realized while the risks are effectively managed.

As we have seen in the case of the British Library and other organizations affected by ransomware, the impacts of an attack can be devastating, but they can also be an opportunity for transformation and growth. By learning from these experiences and applying the lessons of economics, technology, and organizational resilience, we can navigate the future of ransomware with greater confidence and success.

References

[1] British Library. (2024). Cyber Incident Review. https://www.bl.uk/home/british-library-cyber-incident-review-8-march-2024.pdf

[2] Cybersecurity Ventures. (2021). Global Ransomware Damage Costs Predicted To Exceed $265 Billion By 2031. https://cybersecurityventures.com/global-ransomware-damage-costs-predicted-to-reach-250-billion-usd-by-2031/

[3] Kharraz, A., Robertson, W., Balzarotti, D., Bilge, L., & Kirda, E. (2015). Cutting the gordian knot: A look under the hood of ransomware attacks. In International Conference on Detection of Intrusions and Malware, and Vulnerability Assessment (pp. 3-24). Springer, Cham.

[4] Connolly, L. Y., & Wall, D. S. (2019). The rise of crypto-ransomware in a changing cybercrime landscape: Taxonomising countermeasures. Computers & Security, 87, 101568.

[5] Maigida, A. M., Olalere, M., Alhassan, J. K., Chiroma, H., Dada, E. G., & Thakur, S. (2021). Systematic literature review and metadata analysis of ransomware attacks and detection mechanisms. Journal of Reliable Intelligent Environments, 7(4), 269-285.

[6] Bijvank, R., Ferreira, H., & Marin, E. (2022). The role of underground forums in the ransomware-as-a-service business model. Journal of Cybersecurity, 8(1), tyac004.

[7] Alonso, J. A., Turegano, D. M., & Calleja, B. (2023). A deep learning approach for dynamic ransomware prediction and optimal pricing. Journal of Cybersecurity, 9(1), tyad003.

[8] Gonzalez, D., & Hayajneh, T. (2017, December). Detection and prevention of crypto-ransomware. In 2017 IEEE 8th Annual Ubiquitous Computing, Electronics and Mobile Communication Conference (UEMCON) (pp. 472-478). IEEE.

[9] Caulfield, T., & Ioannidis, C. (2022). Dynamic pricing models for ransomware negotiation. Journal of Cybersecurity, 8(1), tyac005.

[10] Berrueta, E., Morato, D., Magaña, E., & Izal, M. (2019). A survey on detection techniques for cryptographic ransomware. IEEE Access, 7, 144925–144944.

[11] Zimba, A., & Chishimba, M. (2021). Ransomware and data protection: a game-theoretical approach. International Journal of Advanced Computer Science and Applications, 12(6), 688–698.

[12] Ibarra, A. B., Cano, J. C., & Venegas, O. (2021). A cyber kill chain model for ransomware attacks. Applied Sciences, 11(12), 5733.

[13] Connolly, L. Y., Wall, D. S., & Lang, M. F. (2022). The economics of cybercrime: The case of ransomware. Journal of Cybersecurity, 8(1), tyac007.

[14] Barlow, T., Becker, B., Morgenstern, J., & Eling, M. (2022). Cyber insurance and ransomware: Modelling the impact of insurance coverage on ransomware attack and defense strategies. Journal of Cybersecurity, 8(1), tyac006.

[15] United States Department of Justice. (2021, June 7). Department of Justice Seizes $2.3 Million in Cryptocurrency Paid to the Ransomware Extortionists Darkside. https://www.justice.gov/opa/pr/department-justice-seizes-23-million-cryptocurrency-paid-ransomware-extortionists-darkside

[16] Turton, W., & Mehrotra, K. (2021, June 4). Hackers Breached Colonial Pipeline Using Compromised Password. Bloomberg. https://www.bloomberg.com/news/articles/2021-06-04/hackers-breached-colonial-pipeline-using-compromised-password

[17] United States Department of the Treasury. (2021, September 21). Treasury Sanctions SUEX OTC S.R.O. for Facilitating Ransomware Transactions. https://home.treasury.gov/news/press-releases/jy0364

[18] The White House. (2021, July 14). Readout of the Second National Cyber Investigative Joint Task Force Summit. https://www.whitehouse.gov/briefing-room/statements-releases/2021/07/14/readout-of-the-second-national-cyber-investigative-joint-task-force-summit/

CHAPTER 17

Case Studies in Confidential Computing

Author:
Ayisha Tabbassum

Overview of Confidential Computing

Introduction to the Concept and Principles of Confidential Computing: Confidential computing addresses a long-standing gap in data security by providing protection not just for data at rest or in transit but also while it's being processed. This novel approach involves securing the computing environment through hardware or software means to prevent unauthorized access to sensitive information. At its core, confidential computing relies on principles like trusted execution environments (TEEs), secure enclaves, and encryption techniques. The goal is to create an environment where sensitive computations can be carried out without exposing the data to the host system, external applications, or malicious actors.

Importance in Today's Data-Driven Environment, Focusing on Protecting Data in Use: In our data-driven world, where analytics, AI, and machine learning are ubiquitous, protecting data throughout its lifecycle

CHAPTER 17   CASE STUDIES IN CONFIDENTIAL COMPUTING

is critical. As computing shifts toward decentralized cloud environments, data security challenges have increased, particularly with processing on shared infrastructure. Confidential computing enables organizations to analyze and utilize sensitive data without fear of exposure. This empowers organizations to use data for research, strategic planning, and innovation while remaining compliant with privacy regulations and maintaining customer trust.

Common Threats and Risks It Aims to Address: Confidential computing targets threats that exploit gaps in data security when sensitive information is actively processed. Malicious insiders, compromised software, and sophisticated external attackers can gain unauthorized access to memory regions where data is temporarily stored during processing. This poses a significant risk to organizations in finance, healthcare, and government that rely on sensitive data. By isolating sensitive operations from the rest of the system, secure enclaves ensure that data remains shielded from unauthorized access, even if parts of the infrastructure are compromised.

Real-World Use Cases and Scenarios

How Different Industries Utilize Confidential Computing to Secure Sensitive Workloads: Various industries use confidential computing to secure highly sensitive workloads. In the financial sector, banks and investment firms leverage confidential computing to process personal financial data, transaction histories, and market analytics while ensuring strict confidentiality. Healthcare institutions rely on it to handle patient data for diagnosis and treatment while remaining compliant with data protection regulations. Similarly, government agencies process classified information with secure enclaves to maintain national security.

Examples Include Finance, Healthcare, and Government Applications: In the financial world, confidential computing enables banks to detect fraud by analyzing customer transactions without exposing private information. Payment processors can aggregate transaction data securely to identify trends that help prevent illegal activity. Healthcare organizations share patient data between hospitals, insurance companies, and researchers while keeping it protected. Government agencies can securely analyze cross-border intelligence to detect threats and develop policies without risking data exposure.

The Impact of Confidential Computing in Improving Compliance and Reducing Risks: Confidential computing minimizes risks by reducing the exposure of sensitive data during computation. It allows organizations to meet the requirements of data protection regulations like GDPR and HIPAA by ensuring that sensitive information is processed in a secure environment. This compliance is crucial for avoiding fines and maintaining customer trust. Furthermore, the ability to analyze sensitive data in a protected manner enhances the accuracy of insights while reducing risks associated with data leaks.

Challenges and Solutions in Implementation

Technical Hurdles Such as Trusted Execution Environments (TEEs) and Their Adoption: Trusted execution environments require specific hardware capabilities to establish isolated processing environments. Not all computing systems are equipped to handle this, limiting the adoption of confidential computing. Additionally, developing applications that operate seamlessly within TEEs demands specialized skills and expertise. This results in a steep learning curve, with developers needing to ensure applications function correctly under such stringent security measures.

Solutions Involving Hardware Isolation, Encryption, and Secure Enclaves: Modern processors increasingly offer built-in support for secure enclaves, allowing isolated regions of memory to be used for confidential computing. Hardware isolation ensures that sensitive workloads remain shielded from the host operating system and other applications.

CHAPTER 17   CASE STUDIES IN CONFIDENTIAL COMPUTING

Encryption methods like homomorphic encryption enable computations on encrypted data without requiring decryption, thus reducing risks. Secure enclaves further bolster this by providing a trusted computing base (TCB) that is resilient against external attacks.

Addressing Challenges like Cost, Performance Overhead, and Usability: Despite significant security benefits, secure enclaves can lead to performance overhead due to the computational cost of encryption and isolation. Furthermore, the usability of confidential computing must be carefully balanced to avoid impacting productivity. Organizations can mitigate these challenges by strategically implementing secure enclaves for the most critical workloads, ensuring that regular operations remain unaffected. Advances in hardware optimization and open-source frameworks will also lower costs over time.

Impact on Data Privacy and Security

How Confidential Computing Enhances Privacy Regulations like GDPR and CCPA: Privacy regulations like GDPR and CCPA impose strict standards on how organizations handle personal data, requiring secure processing environments and transparency. Confidential computing aligns with these regulations by ensuring that sensitive data is only processed within secure enclaves, protecting against unauthorized access. This allows organizations to generate valuable insights from data without compromising privacy.

Role in Secure Data Collaboration and Protecting Intellectual Property: Secure enclaves facilitate the sharing of sensitive data between organizations while maintaining confidentiality. Research institutions can collaborate on sensitive projects without risking data exposure. In manufacturing, intellectual property like blueprints and supply chain data can be processed and shared securely. This fosters innovation and partnerships while protecting proprietary information and maintaining compliance with regulations.

Challenges in Designing and Scaling Secure Solutions: Designing and scaling secure solutions for confidential computing requires a holistic approach to data governance and architecture. Access controls, encryption key management, and secure communication channels are critical components that need to be integrated seamlessly. Scaling presents challenges in terms of maintaining usability and performance across distributed networks while ensuring data integrity.

Lessons Learned and Future Directions

Key Takeaways from Successful Implementation Strategies and Pitfalls to Avoid: Confidential computing should be integrated from the initial design phase for maximum effectiveness. Identifying sensitive workloads early helps determine where secure enclaves should be applied. Thorough testing is essential to prevent performance bottlenecks and compatibility issues. Organizations should avoid relying solely on enclaves without complementing them with comprehensive security measures like encryption and access controls.

How the Confidential Computing Landscape Might Evolve to Meet New Security Challenges: As cyber threats become more sophisticated, confidential computing will likely evolve to provide more robust protections. For instance, future secure enclaves may include built-in AI to detect anomalies and prevent attacks in real time. The growing adoption of confidential computing in cloud environments will also demand interoperability between different platforms, enabling secure data sharing across borders.

The Role of Standards, Collaboration, and Research in Future Evolution: Industry standards like those developed by the Confidential Computing Consortium will play a crucial role in driving the adoption of these technologies. Collaboration between hardware manufacturers, cloud providers, and software developers will streamline implementation

CHAPTER 17 CASE STUDIES IN CONFIDENTIAL COMPUTING

and foster innovation. Research in areas like multi-party computation, homomorphic encryption, and zero-knowledge proofs will further enhance data security, providing organizations with new tools to navigate the evolving cybersecurity landscape.

CHAPTER 18

Case Studies in Cloud Computing

Author:
Ayisha Tabbassum

Evolution of Cloud Computing and Its Security Challenges

The Growth Trajectory of Cloud Computing and Significant Milestones
Cloud computing, with its roots dating back to the early 2000s, has fundamentally transformed how businesses operate. The technology initially gained traction when companies recognized the potential of scalable computing resources that could be provisioned and managed over the Internet. Early milestones include the launch of Amazon Web Services' Elastic Compute Cloud (EC2) in 2006, marking the advent of commercially available infrastructure as a service (IaaS). This innovation was quickly followed by the rise of software-as-a-service (SaaS) platforms such as Salesforce and Google Apps. These milestones reflect a shift in businesses' approach to IT infrastructure, moving away from physical hardware to flexible, on-demand services that empower global collaboration.

How Security Requirements Evolved with Adoption
As cloud computing adoption surged, security concerns emerged due to the unique nature of shared infrastructure. Traditional security

strategies, designed for on-premises systems, proved inadequate for cloud environments. Organizations struggled with securing virtual networks, data storage, and identity management across decentralized platforms. Additionally, compliance with regulations such as GDPR, HIPAA, and CCPA complicated cloud adoption, as these frameworks introduced new data protection obligations. Encryption of data in transit and at rest became a priority to prevent unauthorized access, while identity and access management (IAM) became crucial to ensure that only authorized individuals could interact with sensitive systems.

Challenges Like Multi-cloud Security
The rapid adoption of multi-cloud strategies, where companies use multiple cloud providers to minimize risk and avoid vendor lock-in, has added complexity to cloud security. Each provider operates with different policies, configurations, and features, making it difficult to maintain consistent security across platforms. Data integrity, availability, and privacy can suffer as a result of inadequate visibility and monitoring. Moreover, a lack of standardized protocols and APIs among providers means that security teams must constantly adapt to each vendor's unique environment.

Importance of Shared Responsibility Models
In the cloud, security is a shared responsibility between the cloud provider and the customer. This model means that while providers are responsible for securing the infrastructure (such as physical data centers and hypervisors), customers must secure their applications and data. Misunderstandings of this model have led to incidents where customers mistakenly assumed the provider would secure their data. As a result, companies must clearly understand their obligations, especially regarding data encryption, patch management, and identity controls.

Emerging Solutions and Frameworks
Fortunately, advancements in cloud security frameworks and tools have helped organizations tackle these challenges. Security orchestration, automation, and response (SOAR) systems integrate various tools

and processes to automate threat detection and response. Cloud access security brokers (CASBs) bridge security gaps by providing policy enforcement across multiple platforms. Additionally, zero-trust architecture, which assumes no user or system is inherently trustworthy, offers a more comprehensive framework for safeguarding sensitive assets.

Noteworthy Cloud Security Case Studies

Detailed Analyses of Cloud Security Incidents

One of the most infamous cloud security breaches was the Capital One data breach in 2019, which exposed over 100 million customer records. The breach was caused by a misconfigured firewall in Amazon Web Services (AWS), allowing a hacker to access sensitive information. Despite having security measures in place, the company failed to properly monitor and audit its firewall configurations, highlighting the critical need for visibility into cloud environments.

Success Stories Across Different Organizations

In contrast, Netflix's approach to cloud security is considered exemplary. The streaming giant leverages a "Simian Army" of automated tools that continuously test the resilience of its cloud infrastructure. These tools, which include Chaos Monkey and Security Monkey, proactively seek out vulnerabilities and intentionally disrupt systems to ensure that Netflix's services can withstand attacks or failures. This proactive approach has enabled Netflix to avoid major security incidents while scaling globally.

How Proactive Approaches Ensured Resilience

Proactive strategies, like Netflix's automated testing, have become essential in modern cloud security. By continuously probing for weaknesses and addressing them before they can be exploited, organizations can significantly reduce their risk exposure. Regular security audits, penetration testing, and red team exercises (where internal security teams attempt to breach their own systems) help identify potential vulnerabilities and weaknesses in security protocols.

Reactive Measures Mitigated Breaches

In addition to proactive measures, organizations must have reactive incident response strategies in place to minimize the damage of security breaches. Following the 2017 Equifax breach, which exposed sensitive information of over 140 million customers, the company implemented a comprehensive overhaul of its cybersecurity policies. This included extensive staff training, improved access controls, and a robust incident response plan that helped Equifax recover from the breach and regain public trust.

Common Themes in Case Studies

Across different case studies, some common themes have emerged: the need for comprehensive monitoring and logging, consistent configuration management, and strong identity and access controls. Companies that succeed in securing their cloud environments often have a dedicated security team that continuously reviews their infrastructure for vulnerabilities. Furthermore, they prioritize the automation of routine security tasks to maintain compliance and mitigate risks.

Risk Mitigation Strategies and Best Practices

Multi-factor Authentication

Multi-factor authentication (MFA) is a foundational security practice that requires users to verify their identity using multiple methods before accessing cloud resources. This can include a combination of passwords, one-time verification codes sent via SMS, or biometric data like fingerprint recognition. By implementing MFA, companies can significantly reduce the risk of unauthorized access, especially with the prevalence of phishing attacks and credential theft.

Encryption

Encryption is a critical component of data security, protecting sensitive information from being read by unauthorized parties. In the cloud, data should be encrypted both in transit (as it moves between applications and storage) and at rest (when it is stored in databases or file systems). Modern encryption algorithms like Advanced Encryption Standard (AES) ensure that even if an attacker gains access to encrypted data, it remains unreadable without the corresponding decryption keys.

Network Segmentation

Network segmentation involves dividing an organization's network into smaller, isolated segments to limit the potential spread of cyberattacks. This strategy ensures that if one segment is compromised, the attacker cannot easily access other segments. In cloud environments, network segmentation can be implemented through virtual private networks (VPNs), firewalls, and software-defined networking (SDN) to create secure, isolated networks for different workloads.

Shared Responsibility Models Between Providers and Customers

Effective cloud security relies on a clear understanding of shared responsibility models. Providers handle the security of their infrastructure, but customers must secure their applications, data, and identity systems. This requires customers to implement access controls, manage encryption keys, and monitor their cloud resources. Misconfigurations and unclear division of responsibilities are common causes of cloud security breaches, emphasizing the need for thorough planning and communication.

Best Practices for Cloud Workload Security and Compliance

Maintaining workload security in the cloud requires adherence to best practices like regular vulnerability assessments, automated patch management, and compliance monitoring. Workload configurations should be audited to ensure they align with internal policies and regulatory standards. Additionally, security teams must regularly review user access controls to identify any privileges that may be excessive or outdated. Automated compliance tools can help identify gaps and ensure adherence to frameworks like SOC 2, GDPR, and HIPAA.

Impact of Cloud Adoption on Organizations

Agility, Scalability, and Efficiency
Cloud adoption has fundamentally transformed how organizations operate by offering unprecedented levels of agility, scalability, and efficiency. Companies can now rapidly provision computing resources, enabling them to respond quickly to market changes and customer demands. Scalability allows organizations to handle peak workloads efficiently without incurring the high costs of maintaining excess infrastructure. The cloud also reduces the overhead of managing physical servers and hardware, allowing IT teams to focus on strategic initiatives.

Challenges Encountered in Security Management and Compliance
However, cloud adoption also presents significant security management and compliance challenges. Many organizations struggle to maintain visibility and control over their cloud environments, leading to misconfigurations and data breaches. The complexity of multi-cloud environments exacerbates these challenges, as security teams must juggle different tools, policies, and compliance requirements. Moreover, cloud service providers often have varying data residency requirements, making it difficult to ensure compliance with regional data protection laws.

Changes in Organizational Culture and Roles
Adopting the cloud also necessitates changes in organizational culture and roles. Traditional IT teams accustomed to managing on-premises infrastructure must now adopt a DevOps mindset, focusing on automation, continuous integration, and delivery. Security teams must also evolve to become more proactive, collaborating closely with development teams to embed security early in the software development

lifecycle. This cultural shift requires investment in training and new tools that support collaborative workflows.

Enhancing Collaboration and Innovation
Cloud adoption enhances collaboration across geographically distributed teams. Cloud-based productivity suites like Microsoft 365 and Google Workspace enable seamless document sharing, real-time editing, and communication. In addition, cloud platforms provide access to cutting-edge technologies such as machine learning, data analytics, and IoT, which can be quickly integrated into applications to drive innovation.

Adapting to the Future of Work
The pandemic accelerated the adoption of remote work, and cloud computing became a lifeline for many businesses. Cloud-based virtual desktops, collaboration tools, and secure access solutions have enabled companies to adapt to hybrid and fully remote work models. While this shift has increased employee flexibility and productivity, it has also introduced new security challenges, such as securing remote endpoints and managing third-party access.

Emerging Trends and the Road Ahead

Cloud-Native Security
Cloud-native security is an emerging trend emphasizing security practices that are purpose-built for cloud environments. This approach leverages the inherent scalability and automation capabilities of the cloud to implement security controls directly within the application. For instance, micro-segmentation isolates application components to prevent lateral movement in the event of a breach. Cloud-native security also involves integrating security into the DevOps pipeline, ensuring that applications are continuously tested and monitored for vulnerabilities.

Automation and Zero-Trust Architecture

Automation plays a significant role in modern cloud security by reducing human error and ensuring consistent policy enforcement. Security orchestration tools can automate routine tasks such as compliance checks, access reviews, and incident response. In parallel, zero-trust architecture has gained traction as a comprehensive security framework that verifies every user and system attempting to access a resource. It assumes that no network or user is inherently trustworthy, enforcing stringent access controls and continuous monitoring.

AI/ML in Threat Detection and Management

Artificial intelligence and machine learning (AI/ML) have revolutionized threat detection and security management in the cloud. These technologies can analyze massive datasets to identify unusual patterns indicative of a security incident. For example, anomaly detection algorithms can flag unusual login attempts, data access patterns, or network traffic, allowing security teams to respond more quickly. Additionally, predictive models can identify emerging threats based on historical data, helping organizations proactively defend against potential attacks.

Integration of Security with DevOps

DevOps and security teams have historically operated in silos, leading to security being an afterthought in the software development lifecycle. The integration of security into DevOps workflows, often referred to as DevSecOps, ensures that security controls are embedded early and consistently. This includes automated code scanning, infrastructure as code (IaC) policies, and continuous security testing. DevSecOps also fosters a culture of shared responsibility, where developers are empowered to build secure applications from the outset.

CHAPTER 18   CASE STUDIES IN CLOUD COMPUTING

Edge Computing and Decentralized Security

Edge computing, which brings computation closer to the data source, is poised to redefine cloud security by decentralizing the traditional cloud model. As more devices generate and process data at the edge, organizations must adopt new security strategies that protect these distributed environments. This includes lightweight encryption, device authentication, and real-time anomaly detection. Edge computing will also require tighter integration between cloud and edge security platforms, ensuring seamless data protection across hybrid architectures.

Conclusion

The evolution of cloud computing has dramatically transformed organizations, offering unparalleled opportunities for growth and innovation. However, this transformation has also introduced new security challenges that require organizations to adopt proactive, collaborative, and adaptive strategies. From multi-factor authentication and encryption to zero-trust architecture and cloud-native security, best practices continue to evolve to meet emerging threats.

Case studies of cloud security incidents and successes underscore the importance of a well-defined shared responsibility model and the need for comprehensive monitoring, automated compliance, and consistent configuration management. The rise of DevSecOps and AI/ML in threat detection demonstrates how security practices must be embedded throughout the development lifecycle.

Looking ahead, emerging trends like edge computing and decentralized security will redefine the way organizations protect their data and applications. Ultimately, cloud security will continue to evolve, emphasizing adaptability, collaboration, and continuous improvement to ensure resilience in the face of an ever-changing threat landscape.

CHAPTER 18 CASE STUDIES IN CLOUD COMPUTING

Multiple Choice Questions

1. What milestone marked the beginning of commercially available Infrastructure as a Service (IaaS)?

    A) Launch of Microsoft Azure

    B) Launch of Google Cloud Platform

    C) Launch of Salesforce

    D) Launch of Amazon Web Services EC2

2. Which of the following challenges is associated with implementing multi-cloud security?

    A) Managing physical server hardware

    B) Maintaining a shared responsibility model

    C) Ensuring consistent security across different platforms

    D) Creating secure on-premises networks

3. In the shared responsibility model, what are customers primarily responsible for securing?

    A) Cloud infrastructure and physical data centers

    B) Applications, data, and identity systems

    C) Trusted Execution Environments (TEEs)

    D) Network segmentation

4. Which proactive testing strategy does Netflix employ to ensure its cloud services are resilient to attacks or failures?

    A) Simian Army Tools

    B) Regular Staff Training

CHAPTER 18   CASE STUDIES IN CLOUD COMPUTING

   C) Incident Response Plans

   D) Penetration Testing

5. What common security challenge did the Capital One breach in 2019 highlight?

   A) Lack of firewall monitoring and configuration

   B) Mismanaged encryption keys

   C) Excessive network segmentation

   D) Absence of AI-driven monitoring

6. How does encryption help protect sensitive data in the cloud?

   A) By reducing network traffic

   B) By securing data in transit and at rest

   C) By blocking phishing attempts

   D) By automating compliance

7. What role does network segmentation play in cloud security?

   A) Isolating different parts of the network to prevent lateral movement in case of a breach

   B) Restricting network traffic to known endpoints

   C) Providing dynamic scaling and resource allocation

   D) Allowing for easy user authentication and authorization

8. Which security tool integrates various processes to automate threat detection and response?

   A) Cloud-Native Security

   B) Zero-Trust Architecture

CHAPTER 18   CASE STUDIES IN CLOUD COMPUTING

    C) Security Orchestration, Automation, and Response (SOAR)

    D) Public Key Infrastructure (PKI)

9. Which of the following statements best describes zero-trust architecture?

    A) It assumes internal networks are secure

    B) It automatically approves trusted systems

    C) It provides access based on device authentication only

    D) It assumes no network or system is inherently trustworthy

10. In the shared responsibility model, what are cloud service providers responsible for securing?

    A) Encryption key management

    B) Data stored in the cloud

    C) The physical infrastructure and hypervisors

    D) Cloud-based applications

11. How has cloud adoption transformed organizational agility?

    A) By increasing reliance on physical servers

    B) By creating region-specific regulatory frameworks

    C) By enabling rapid provisioning of computing resources

    D) By reducing shared responsibility

12. What is the primary focus of cloud-native security practices?

    A) Automating incident response using SOAR tools

    B) Implementing security controls tailored to the inherent scalability and automation capabilities of the cloud

CHAPTER 18 CASE STUDIES IN CLOUD COMPUTING

    C) Securing on-premises network traffic

    D) Training staff on endpoint management

13. Which encryption standard is commonly used to protect data in transit and at rest in cloud environments?

    A) RSA

    B) MD5

    C) SHA-1

    D) Advanced Encryption Standard (AES)

14. What is the goal of DevSecOps?

    A) Securing software after it is deployed

    B) Isolating cloud networks into multiple segments

    C) Embedding security early and consistently into DevOps workflows

    D) Creating redundancy for cloud workloads

15. Which cloud security solution helps enforce policies across multiple cloud platforms?

    A) Security Information and Event Management (SIEM)

    B) Cloud Access Security Brokers (CASBs)

    C) Security Operations Centers (SOCs)

    D) Trusted Execution Environments (TEEs)

16. Which of the following challenges organizations in securing their multi-cloud environments?

    A) Limited collaboration tools

    B) Inadequate staff training

C) Varying tools, policies, and compliance requirements across cloud service providers

D) Scarcity of automation tools

17. Why is AI/ML valuable in cloud security management?

    A) It reduces overall cloud computing costs

    B) It eliminates the need for traditional firewalls

    C) It can identify unusual patterns indicative of a security incident

    D) It provides unlimited computing resources

18. What is a key feature of the zero-trust security model?

    A) Trusting users once authenticated

    B) Assuming that no network or user is inherently trustworthy

    C) Allowing unauthenticated internal users to access sensitive data

    D) Providing access based solely on user role

19. Which key takeaway can be drawn from successful cloud security case studies?

    A) Proactive measures are generally unnecessary

    B) Consistent configuration management is critical

    C) Shared responsibility models should not be enforced

    D) Network segmentation is redundant in the cloud

CHAPTER 18 CASE STUDIES IN CLOUD COMPUTING

20. What is one of the main advantages of edge computing?

    A) Increased reliance on physical servers

    B) Standardized configurations across providers

    C) Decentralized processing closer to the data source

    D) Simplified data encryption

21. What is a significant outcome of integrating security into the DevOps pipeline?

    A) Reduced staff involvement in testing

    B) Fewer incidents of unauthorized access

    C) Elimination of shared responsibility

    D) Increased security after software deployment

22. What should security teams regularly review to minimize security risks?

    A) Physical server logs

    B) User access controls for excessive or outdated privileges

    C) Public cloud infrastructure code

    D) Multi-cloud network traffic logs

23. What is a key security benefit of implementing multi-factor authentication (MFA)?

    A) Automating compliance audits

    B) Reducing the need for encryption

    C) Minimizing the risk of unauthorized access

    D) Simplifying data backup processes

CHAPTER 18 CASE STUDIES IN CLOUD COMPUTING

24. What role do predictive models play in cloud security?

    A) Identifying patterns indicative of a security incident

    B) Generating compliance certificates

    C) Monitoring user behavior in internal networks

    D) Encrypting sensitive data

25. What emerging trend emphasizes security practices purpose-built for cloud environments?

    A) Zero-Trust Architecture

    B) Multi-Factor Authentication

    C) Cloud-Native Security

    D) Security Information and Event Management (SIEM)

Answers

1. D
2. C
3. B
4. A
5. A
6. B
7. A
8. C
9. D

CHAPTER 18 CASE STUDIES IN CLOUD COMPUTING

10. C
11. C
12. B
13. D
14. C
15. B
16. C
17. C
18. B
19. B
20. C
21. B
22. B
23. C
24. A
25. C

CHAPTER 19

Case Studies in Enterprise Security Architecture

Author:

Ayisha Tabbassum

Enterprise security architecture is the practice of designing, implementing, and maintaining a holistic and coherent security posture for an organization. It involves aligning security objectives with business goals, applying security principles and best practices across all layers of the enterprise, and adapting to changing requirements and threats. In this chapter, we will explore the concepts and components of modern enterprise security architecture, and how different frameworks can help guide security practitioners in developing secure solutions. We will also look at some real-world examples from various industries that showcase how effective security architectures can protect critical assets and enable business value. Finally, we will discuss the challenges and strategies of balancing usability and security in enterprise settings and provide some key takeaways and recommendations for practitioners.

CHAPTER 19   CASE STUDIES IN ENTERPRISE SECURITY ARCHITECTURE

Understanding Enterprise Security Architecture

Enterprise security architecture is not a one-size-fits-all solution, but rather a customized approach that considers the specific needs and characteristics of each organization. However, there are some common elements and definitions that can help us understand the scope and purpose of enterprise security architecture. According to the Open Group, a global consortium that develops open standards and frameworks, enterprise security architecture is defined as

> A subset of the enterprise architecture that provides a strategic context for the deployment of IT security capabilities, based on a clear understanding of the threats, vulnerabilities, impacts, and risks to the enterprise's assets and missions.

This definition highlights several important aspects of enterprise security architecture, such as:

> It is a subset of the enterprise architecture, which is a broader term that encompasses the entire structure and operation of an organization, including its business, information, application, technology, and security domains.

> It provides a strategic context, which means that it aligns the security vision and goals with the overall business strategy and objectives, and defines the roles and responsibilities of the stakeholders involved.

It is based on a clear understanding of the threats, vulnerabilities, impacts, and risks, which means that it conducts a comprehensive and systematic analysis of the potential sources and consequences of security breaches, and prioritizes the protection of the most critical and valuable assets and missions.

It deploys IT security capabilities, which means that it implements and integrates the appropriate security controls and mechanisms across the enterprise, such as policies, processes, tools, technologies, and people.

Based on this definition, we can identify some of the key components of enterprise security architecture, such as

Security principles and standards, which are the high level guidelines and rules that govern the security behavior and decisions of the organization.

Security models and patterns, which are the abstract representations and templates that describe the security properties and relationships of the enterprise elements.

Security services and functions, which are the specific capabilities and features that provide security functionality and support to the enterprise.

Security components and technologies, which are the concrete implementations and instances of the security services and functions.

CHAPTER 19 CASE STUDIES IN ENTERPRISE SECURITY ARCHITECTURE

The Open Group Architecture Framework (TOGAF), which is a comprehensive and generic framework for developing and managing enterprise architectures, including security architectures. It provides a set of methods, processes, tools, and best practices that cover the entire lifecycle of architecture development, from vision and scope to implementation and governance.

The Sherwood Applied Business Security Architecture (SABSA), which is a specific framework for developing and managing enterprise security architectures. It is based on the concept of a "security service oriented architecture," which means that it focuses on delivering security services that meet the business needs and expectations of the stakeholders. It uses a layered model that covers six aspects of security architecture: context, business, information, application, technology, and physical.

The Committee on National Security Systems (CNSS) Security Architecture Framework, which is a specific framework for developing and managing enterprise security architectures in the US federal government sector. It is based on the concept of a "system of systems," which means that it views the enterprise as a collection of interconnected and interdependent systems that share common security objectives and requirements. It uses a three tiered model that covers the strategic, operational, and tactical levels of security architecture.

CHAPTER 19   CASE STUDIES IN ENTERPRISE SECURITY ARCHITECTURE

Designing and Implementing Secure Enterprise Solutions

Using these frameworks as guidance, security practitioners can design and implement secure enterprise solutions that align with the business objectives and address the security challenges of the organization. The exact steps and methods may vary depending on the framework and the context, but some of the common activities and considerations are as follows:

> Establishing the security architecture vision and scope, which involves defining the purpose, scope, and boundaries of the security architecture and identifying the key stakeholders, drivers, and objectives.

> Conducting the security architecture analysis, which involves collecting and analyzing the relevant information about the current state and the desired state of the enterprise, such as its assets, missions, functions, processes, systems, technologies, threats, vulnerabilities, risks, and controls.

> Developing the security architecture design, which involves creating and documenting the logical and physical models and patterns of the security architecture, such as its principles, standards, services, functions, components, and technologies.

> Implementing the security architecture solution, which involves deploying and integrating the security architecture components and technologies into the enterprise environment, and testing and verifying their functionality and performance.

CHAPTER 19   CASE STUDIES IN ENTERPRISE SECURITY ARCHITECTURE

Managing and governing the security architecture lifecycle, which involves monitoring and evaluating the security architecture solution, and maintaining and updating it to reflect the changes in the enterprise and the security landscape.

Throughout these activities, security practitioners should consider the following aspects of secure enterprise solutions:

Secure network design, which involves designing and implementing a network infrastructure that supports the security objectives and requirements of the enterprise, such as confidentiality, integrity, availability, authentication, authorization, and accountability. Some of the key elements of secure network design are: network segmentation, which divides the network into smaller and isolated zones based on their security levels and functions; network access control, which regulates the access to the network resources and services based on the identity and attributes of the users and devices; network encryption, which protects the data in transit from unauthorized interception and modification; network monitoring, which detects and alerts on the network anomalies and incidents; and network resilience, which ensures the continuity and recovery of the network operations in the event of disruptions or failures.

Secure access control, which involves designing and implementing a system that grants or denies access to the enterprise resources and services based on the security policies and rules of the organization. Some of the key elements of

CHAPTER 19 CASE STUDIES IN ENTERPRISE SECURITY ARCHITECTURE

secure access control are: identity and credential management, which establishes and manages the identity and credentials of the users and devices that access the enterprise; authentication, which verifies the identity and credentials of the users and devices that access the enterprise; authorization, which determines the level and scope of access that the users and devices have to the enterprise resources and services; and auditing, which records and reviews the access activities and events for compliance and accountability purposes.

Secure data security, which involves designing and implementing a system that protects the data of the enterprise from unauthorized access, use, disclosure, modification, or destruction. Some of the key elements of secure data security are: data classification, which categorizes the data based on its sensitivity and value; data encryption, which protects the data at rest and in transit from unauthorized access and modification; data masking, which obscures the sensitive data from unauthorized viewers; data backup, which creates and maintains copies of the data for recovery purposes; and data disposal, which destroys the data when it is no longer needed.

Secure compliance, which involves designing and implementing a system that ensures that the enterprise meets the legal, regulatory, and contractual obligations and standards related to security. Some of the key elements of secure compliance are: security policy, which defines the security goals, roles, responsibilities, and rules of the organization; security governance, which oversees and coordinates the security activities and initiatives of the organization; security assessment, which evaluates and measures the security posture and performance of the organization; security audit, which verifies and validates the compliance of the organization with the security requirements and standards; and security reporting, which communicates and demonstrates the security status and achievements of the organization.

Case Studies Showcasing Successful Security Architectures

To illustrate how enterprise security architecture can be applied in practice, we will present some case studies from different industries that showcase how successful security architectures have protected critical assets and enabled business value. These case studies are not meant to be exhaustive or definitive, but rather to provide some examples and insights into the challenges and strategies of enterprise security architecture in real world scenarios.

Case Study 1: Banking Sector

The banking sector is one of the most regulated and targeted industries in terms of security, as it deals with sensitive and valuable data and transactions, such as personal and financial information, money transfers, and online payments. The banking sector faces various security threats, such as cyberattacks, fraud, identity theft, money laundering, and data breaches,

CHAPTER 19   CASE STUDIES IN ENTERPRISE SECURITY ARCHITECTURE

which can result in significant financial losses, reputation damage, customer dissatisfaction, and legal penalties. Therefore, the banking sector needs to implement robust and comprehensive security architectures that can protect its assets and missions, and comply with the security standards and regulations, such as the Payment Card Industry Data Security Standard (PCI DSS), the Gramm Leach Bliley Act (GLBA), and the Basel III Accord.

One example of a bank that has implemented a successful security architecture is Bank of America, one of the largest and most diversified financial institutions in the world. Bank of America has developed and deployed a security architecture that is based on the following principles and practices:

> Security by design, which means that security is embedded and integrated into the business processes and systems from the inception to the retirement stages, and that security requirements and controls are defined and implemented upfront, rather than as an afterthought or a patchwork.

> Security defense in depth, which means that security is applied at multiple layers and dimensions of the enterprise, such as the network, the application, the data, the user, and the physical levels, and that security controls are diversified and redundant, rather than relying on a single point of failure or a single line of defense.

> Security risk management, which means that security is driven and guided by a systematic and proactive process of identifying, assessing, prioritizing, mitigating, and monitoring the security risks to the enterprise, and that security decisions and investments are based on the risk appetite and tolerance of the organization.

CHAPTER 19 CASE STUDIES IN ENTERPRISE SECURITY ARCHITECTURE

> Security intelligence and analytics, which means that security is supported and enhanced by a data-driven and evidence-based approach of collecting, analyzing, and interpreting the security data and information, and that security insights and actions are derived from the security metrics and indicators.
>
> Security awareness and culture, which means that security is shared and promoted as a collective and collaborative responsibility and value of the organization, and that security education and training are provided and encouraged for all the stakeholders, including the employees, the customers, and the partners.

By adopting these principles and practices, Bank of America has achieved a security architecture that is aligned with its business strategy and objectives, and that is resilient and adaptable to the changing security landscape and demands. Some of the benefits and outcomes of its security architecture are

> Enhanced security posture and performance, which means that the bank has reduced its exposure and vulnerability to security threats and increased its detection and response capabilities and efficiency
>
> Improved customer trust and satisfaction, which means that the bank has maintained and strengthened its reputation and relationship with its customers, and provided them with secure and convenient services and products
>
> Increased compliance and governance, which means that the bank has met and exceeded the security standards and regulations and demonstrated its accountability and transparency to the regulators and the public

CHAPTER 19 CASE STUDIES IN ENTERPRISE SECURITY ARCHITECTURE

Multiple Choice Questions

1. What does enterprise security architecture primarily align with?

   A) Technical goals

   B) Security best practices

   C) Business goals

   D) International standards

2. Which of the following best defines enterprise security architecture?

   A) Security rules applied uniformly

   B) Holistic approach aligning security with business strategy

   C) A singular solution for all organizations

   D) Compliance with security frameworks

3. Which framework emphasizes security controls aligned with business needs, providing comprehensive strategies to ensure secure information flows?

   A) TOGAF

   B) COBIT

   C) NIST

   D) ISO 27001

4. Which framework adopts a six layered model, focusing on delivering security services that meet stakeholders' business needs?

   A) TOGAF

   B) SABSA

   C) ISO 27001

   D) NIST

5. What is the primary purpose of a zero trust architecture?

   A) Network encryption

   B) Unrestricted data sharing

   C) Continuous trust evaluation

   D) Simplified security policies

6. Which element is not part of secure network design?

   A) Network segmentation

   B) Network access control

   C) Network monitoring

   D) Data classification

7. What does "defense in depth" imply in security architecture?

   A) Relying on perimeter defense

   B) Applying security across multiple layers

   C) Standardizing all access policies

   D) Having a single comprehensive control

CHAPTER 19 CASE STUDIES IN ENTERPRISE SECURITY ARCHITECTURE

8. What is the role of identity and credential management in secure access control?

   A) Managing data backups

   B) Establishing users' identities and credentials

   C) Classifying sensitive data

   D) Encrypting network traffic

9. What is the primary objective of security compliance?

   A) Increasing customer satisfaction

   B) Simplifying network design

   C) Meeting legal, regulatory, and contractual obligations

   D) Enhancing data encryption

10. Which of these activities is not part of the security architecture lifecycle?

    A) Architecture analysis

    B) Developing design patterns

    C) Setting up backups

    D) Managing architecture updates

11. What strategy did Bank of America implement to ensure security by design?

    A) Integrated security into all system stages

    B) Installed SIEM systems across branches

    C) Established biometric identification

    D) Allowed access through virtual networks

CHAPTER 19 CASE STUDIES IN ENTERPRISE SECURITY ARCHITECTURE

12. Which of these is a benefit achieved through Bank of America's security architecture?

    A) Enhanced security posture

    B) Improved data classification

    C) Reduced encryption needs

    D) Increased automation

13. Which standard specifically relates to the security of card payments in the banking industry?

    A) HIPAA

    B) PCI DSS

    C) ISO 27001

    D) COBIT

14. Which of these is a key takeaway from security architecture in the healthcare sector?

    A) Micro segmentation provides secure data isolation

    B) Regular audits are unnecessary

    C) Patient data does not need encryption

    D) Firewalls replace monitoring systems

15. Which aspect of Bank of America's security architecture deals with "evidence-based data analysis"?

    A) Network segmentation

    B) Identity management

    C) Security intelligence and analytics

    D) Automated incident response

CHAPTER 19 CASE STUDIES IN ENTERPRISE SECURITY ARCHITECTURE

16. What is the role of network encryption in secure enterprise architecture?

    A) Protects data in transit

    B) Enhances physical security

    C) Simplifies access control

    D) Automates compliance reports

17. Which aspect of secure access control grants or denies access to enterprise resources?

    A) Authentication

    B) Authorization

    C) Network segmentation

    D) Data masking

18. Why is secure backup essential for data security?

    A) It helps identify potential vulnerabilities

    B) It classifies data into various levels

    C) It provides copies of data for recovery purposes

    D) It automates security reports

19. What concept does SABSA emphasize in its framework for security architecture?

    A) Strategic network architecture

    B) Role-based monitoring

    C) Security service-oriented architecture

    D) Centralized user management

CHAPTER 19 CASE STUDIES IN ENTERPRISE SECURITY ARCHITECTURE

20. Which of the following is an element of secure compliance?

    A) Security assessment

    B) Threat response

    C) Network architecture

    D) Data disposal

21. Which concept ensures that security controls are appropriate for a user's role or function?

    A) Network encryption

    B) Least privilege

    C) Zero trust

    D) Threat modeling

22. What practice aligns with the idea of "security risk management" at Bank of America?

    A) A reactive response to evolving threats

    B) Managing security risks based on business tolerance

    C) Limiting the organization's spending on security

    D) Creating data masks for sensitive information

23. Which architecture pattern involves analyzing the security impacts of interconnected systems?

    A) Network segmentation

    B) System of systems

    C) Role-based access

    D) Distributed computing

CHAPTER 19 CASE STUDIES IN ENTERPRISE SECURITY ARCHITECTURE

24. How can "security awareness and culture" improve an organization's security architecture?

    A) By building collective responsibility and education

    B) By reducing technical vulnerabilities

    C) By limiting user access

    D) By developing internal reporting systems

25. Which of the following best describes "security governance"?

    A) Aligning security goals with regulatory obligations

    B) Overseeing and coordinating security activities

    C) Setting up network perimeter security

    D) Managing IT staff rotations

Answers

1. C
2. B
3. A
4. B
5. C
6. D
7. B
8. B
9. C

CHAPTER 19   CASE STUDIES IN ENTERPRISE SECURITY ARCHITECTURE

10. C
11. A
12. A
13. B
14. A
15. C
16. A
17. B
18. C
19. C
20. A
21. B
22. B
23. B
24. A
25. B

CHAPTER 20

Case Studies in Energy

Author:
Anirudh Khanna

Overview of Ransomware Attacks in the Energy Industry

The trend of digitalizing energy installations and power grids globally is growing to improve control, security, and operational efficiency. According to a report published by the International Energy Agency [31], the number of intelligent and connected devices on the global energy and power grids will rise to 30 and 40 billion by 2025. However, smart grids, which are electrical grids that rely on digital systems for control and monitoring [43], alongside energy systems operated and monitored using online and digital systems, are more vulnerable to cyberattacks as witnessed in the EU, where ransomware attacks targeting energy installations in 2022 stood at 21 cases by October of the same year [33]. Ransomware attacks on energy installations such as nuclear power plants, wind turbine farms, and hydroelectric power grids are currently one of the most concerning trends in cybersecurity.

Therefore, the global energy industry is increasingly vulnerable to cybercrime, as demonstrated in a report by Allianz [32], which stated a

CHAPTER 20   CASE STUDIES IN ENERGY

cyberattack using ransomware or other methods on the US power grid would result in economic costs between 240 billion and 1 trillion USD. These financial costs would emanate from downtime, insurance claims, and other factors, further compounding the vulnerability of American and global energy installations. Unfortunately, modern cybersecurity capabilities are not adopted to prevent threats like ransomware attacks in the energy industry because of the high implementation costs. In 2021, security rating services provider Black Kite [34] reported that the oil energy sub-sector was 28% susceptible to ransomware attacks, while the entire energy sub-sector was 25% susceptible to similar attacks.

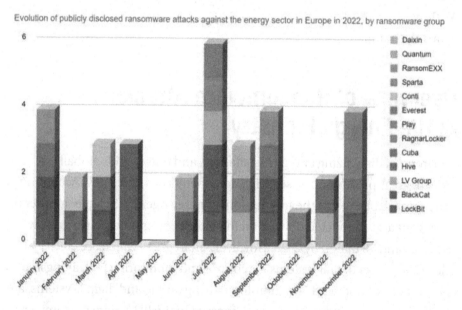

Figure 20-1. *Ransomware Attacks Affecting the EU Energy Sector in 2022 According to Cybercrime Group [35]*

The energy industry is beautiful for perpetrators of ransomware attacks because of the critical nature of power grids, the safety risk of nuclear power reactors, and the economic and geopolitical nature of utilities [7]. Although most ransomware attacks targeting the energy sector, like Mazewalker

CHAPTER 20 CASE STUDIES IN ENERGY

and Ryuk, sought financial gains, emerging trends aimed at infrastructural damage exist. The Triton and DarkSide malware used recently in Saudi Arabia and the United States targeted safety control systems seeking to compromise site safety. In contrast, Stuxnet sought to damage energy installations by using electrical fires and physically malfunctioning the system. These are demonstrations of the changing operational terrain of ransomware attacks targeting energy installations.

Selected Case Studies of Ransomware Attacks in the Energy Industry

Case Study 1: Nordex Group SE and Deutsche Windtechnik Ransomware Attacks

Introduction to the Case

Nordex Group SE and Deutsche Windtechnik are major energy industry players in Germany, with the first firm positioned as a wind turbine manufacturer and the latter providing maintenance services. The Nordex Group SE is a global leader in designing and installing wind turbines, which means it handles many contracts in the EU, North America, and Asia. On March 31, 2022, the company's IT personnel noted the initial stages of a well-designed ransomware attack when there are signs of brute force attacks or malware [44].

These operational changes are characterized by challenges accessing the remote-control stations used to control wind turbines remotely [2]. According to Nordex Group SE's cybersecurity crisis management protocols, IT personnel immediately shut down all remote access points between wind turbine farms and remote access locations. The firm's response strategy aligns with the ransomware recovery framework protocol of reducing access to data and systems after initial detection.

405

CHAPTER 20   CASE STUDIES IN ENERGY

Afterward, the Deutsche Windtechnik Company was targeted on April 11, 2022, perhaps in retaliation for the response to Nordex Group SE's fast action of reducing access by shutting down remote access to the wind turbine farms [2]. Notably, both Nordex Group SE and Deutsche Windtechnik operate on many contracts concurrently, with the former company designing and installing wind turbines, while the latter offers operational and maintenance support. Deutsche Windtechnik's ransomware attack was more detailed because the malware turned off control and remote connectivity to more than 2000 wind turbines in Germany.

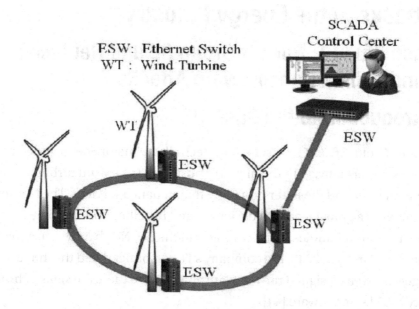

Figure 20-2. General Windfarm Communications and Control Infrastructure Showing Fan Area Network (FAN) [4]. Each turbine has a switch embedded in its base to a SCADA ethernet network, the design of which depends on the turbine designer and installer. For instance, Siemens uses the PROFINET protocol, while General Electric relies on OPC or ModBus protocols. The switches are remotely connected to engineering control and monitoring stations using VPN tunnels and industrial ethernet.

Detailed Analysis

The main access point for ransomware attacks targeting wind turbine farms is maintenance personnel who access the control centers or actual turbines using VPN tunnels and SCADA systems [7]. Poor password management is a common access point for initial access brokers (IABs) or attackers who take over control of the industrial networks, controlling turbines and the main control centers. During the ransomware analysis stages of the companies' ransomware recovery process, the IT personnel at the Nordex Group SE and Deutsche Windtechnik firms discovered Conti malware code evidence [8]. The Conti Malware Group was a renowned cybercrime group responsible for healthcare system attacks in Ireland and government system compromise in Costa Rica. The group was suspected to have compromised more than 650GB of data belonging to the Costa Rican government because the 10 million USD ransom was not paid before disappearing in June 2022. Although the government declared a state of emergency citing cyber-terrorist attacks, no action was taken to remedy the attack, demonstrating Latin America's critical infrastructure vulnerability to ransomware [2]. Also, the malware seemed designed to infiltrate data and victim systems that relied on remote access points controlling more extensive infrastructures and data centers.

Responses and Challenges to Their Implementation

The ransomware attacks on both energy companies were resolved within weeks, which was fortunate considering that the norm for similar cyber threat scenarios is months, as evidenced by the earlier Enercon attack. The primary response by both companies' IT teams was to shut down remote access to remote energy installations, which denied the attackers further access to data and critical systems, reducing the attack's integrity [9]. However, several observations pointed to challenges in future attacks on similar installations.

Notably, the ransomware attack revealed two critical vulnerabilities in its victims' systems and data that contributed to their attack. These vulnerabilities also denoted challenges to the initial response to the ransomware attacks. First, there were many access points between turbine farms and remote-control access points with poor cybersecurity design [35]. Second, the cybersecurity strategies in place at both companies lacked enough redundancies to enable almost immediate reconfiguration and recovery, which is critical in energy sector infrastructure at power generation levels of operation [10]. Both firms' operations and control systems seemed insufficiently segmented from their IT networks to prevent a wide 'blast radius' in case of a ransomware attack. The sheer scale of this attack also demonstrated how small the segmentation, as mentioned earlier, was because Deutsche Windtechnik operates more than 7500 turbines in Germany, and more than 2000 were affected [2].

Effectiveness of the Strategies

The company's IT personnel's decision to immediately shut down all remote access points between wind turbine farms and remote access locations was highly influential in compromising the ransomware attack. This response strategy by the firm under attack aligns with the ransomware recovery framework protocol of reducing access to data and systems after initial detection. Unlike other energy installations like ENERCON, which took months to recover due to lags in reducing the attackers' access to IT infrastructure, the German teams acted swiftly, thwarting the attack in good time.

CHAPTER 20   CASE STUDIES IN ENERGY

Case Study 2: Saudi Aramco's Petro Rabigh Complex Ransomware Attack
Introduction to the Case

On August 04, 2017, emergency alarms activated at Saudi Aramco's Petro Rabigh Complex, resulting in a slow shut-off of critical Triconex safety systems [3]. However, the maintenance engineers on duty should have noticed something visually while inspecting the remotely operated switches. However, rapid increases in operating pressures, temperatures, and leaks in the piping handling explosive and toxic by-product gases alerted engineers to the possibility of a remotely orchestrated intrusion into their network. The monitoring engineers also failed to notice anything outside operational parameters on their monitoring computer terminals because the safety system activation was part of a well-orchestrated ransomware attack using Triton malware. Triconex safety systems are used worldwide in petrochemical installations to control the flow of various chemicals like hydrogen sulfide, methane, and multiple fuels [13] [15]. Notably, the choice of safety systems activated was tactical because it only targeted systems within the 3000-acre petrochemical complex that dealt with poisonous and dangerous petrochemical compounds [3].

Investigator detected the ransomware attack soon after and initiated an immediate response, which, according to the ransomware recovery framework, involves stopping further access to data and system infrastructure. Afterward, the ransomware analysis process yielded evidence of code related to the Triton malware, similar to Stuxnet, used previously to compromise Iran's nuclear centrifuges [16]. Incidentally, the ransomware was advanced enough to compromise several levels of the highly risky petrochemical complex through pressure alterations and valve manipulation, releasing poisonous hydrogen sulfide gases in millions of liters and explosions of methane and other fuels in pressure containers and piping systems.

CHAPTER 20   CASE STUDIES IN ENERGY

Detailed Analysis

The increased digitalization of extensive energy infrastructure with machine learning capabilities has placed them at risk of ransomware, among other cyber threats. Schneider Electric is a principal designer and installer of remote access and control equipment for petrochemical installations. Initially, the Schneider Electric range of Triconex safety control equipment was designed to continue functioning even during unauthorized remote access to ensure the operational safety of the industrial processes, resulting in a significant cybersecurity design flaw [36]. Conversely, the control equipment constantly scans for unsafe operating conditions using software and communicates with remote controllers and operating stations using ethernet resources [3]. If the system detects a minimum number of unsafe parameters, it operates a shutdown autonomously without human intervention, thus making it ideal for such high-risk industrial operations. However, the digital gap between human operators monitoring the system using ethernet resources like SCADA or other systems and its physical location leaves it vulnerable to cyberattacks [17]. The ransomware on the Petro Rabigh Complex in Saudi Arabia evidenced this weakness.

Response and Challenges to Their Implementation

Saudi Aramco's executive management and IT team decided to contract several investigative teams to unravel the ransomware attack. Part of the reasoning for such a multi-pronged approach toward sorting the malware attack was to understand the nature of the complex cyberattack and improve the oil company's preparedness, considering that petrochemical installations were fast becoming likely targets of ransomware attacks intended to cause catastrophic failure alongside financial gain for the cyber criminals [14].

One of the challenges faced during the response was the sheer size of the petrochemical installation at Petro Rabigh Complex. The installation sits on over 3000 acres, meaning physical inspection to complement digital monitoring was impossible in a timely fashion due to the physical dimensions of petrochemical sites [37]. Therefore, once the safety control system was compromised by ransomware, the IT team had little time to detect this attack and affect the ransomware recovery framework protocols, beginning with controlling the extent of the attack by malware. Second, it was evident that the IT system at Petro Rabigh Complex's control stations lacked enough redundancy to ensure that backups could be restarted immediately after infected systems and data were infected, which was a severe cybersecurity design flaw [38]. Unfortunately, many industrial and energy installations lack enough cybersecurity system redundancy to ensure smooth restart using backups once infected systems and data are isolated for remediation. Finally, the Saudi oil industry is heavily guarded from complete shutdown due to the importance of petrochemicals to the country's economy [11]. Therefore, introducing several IT consultants concurrently after the attack may have hindered the realization of a hasty solution because of interference among these investigative and forensic IT professionals from France and the United States.

Effectiveness of Strategies and Lessons Learned

The ransomware attack at Petro Rabigh Complex was resolved within two months, although the details of the investigative reports remained discrete. Notably, the IT team's strategy to isolate infected data and systems effectively reduced the attack's intended scope of operations and effect [18]. Some lessons learned after the remediation of the Saudi Aramco ransomware attack touched on espionage. While most ransomware attacks were financially motivated, foreign spy agencies were weaponizing such malware and their designers against American, European, and Middle Eastern energy installations intent on catastrophic failure [12].

CHAPTER 20 CASE STUDIES IN ENERGY

American researchers had uncovered many attacks from China and Russia intent on its energy infrastructure with little interest in ransom, meaning the focus was on economic sabotage and humanitarian calamity. Therefore, ransomware attacks on energy infrastructure assumed a global security perspective because criminals collaborated with spy agencies to weaponize large energy installations containing highly poisonous or risky materials like nuclear fuel and petrochemicals.

Case Study 3: Colonial Oil Pipeline Attack in the United States
Introduction to the Case

The Colonial Oil Pipeline is one of the largest oil pipeline systems in the world, running more than 5000 miles from Texas to New Jersey [20]. This energy distribution system is critical to the economy of the United States and several other states because it supplies almost 45% of all the fuel products on the East Coast. Hence, the energy installation is heavily digitalized with numerous redundancies and monitoring stations along its entire length, checking performance, operational safety, and IT-based risk [19].
Therefore, when the oil pipeline's digital access and control systems shut down for several days in May 2021, it was evident that a ransomware attack had begun [20]. Notably, the attacker did not compromise the digital systems that control the flow of oil and fuel products but had complete control of digital access to the system. This ransomware attack was executed by a cybercrime group called DarkSide.

CHAPTER 20   CASE STUDIES IN ENERGY

Detailed Analysis

The Colonial Oil Pipeline is headquartered in Georgia's city of Alpharetta, where its Chief Technology Officer and other chief executive officers reside. Unfortunately, such geographical challenges and the bureaucratic nature of such federal industrial sites deter speedy cybersecurity responses [39]. Notably, Colonial Pipeline has numerous control and monitoring stations along its entire length to New Jersey on the US East Coast. DarkSide's ransomware attack was modernized because it did not compromise the oil pipeline's flow of energy products like heating gas, motor fuels, and natural gas to negate the need for negotiation [40]. It took control of essential digital systems and data in accounting and billing departments amounting to 100GB [20].

The ransomware attack strategy implemented by DarkSide was to take control of essential data and system functionality while allowing the energy utility to continue operations. Also, the attackers took control of the numerous VPN-based access points control and monitoring personnel working at various control stations used to gain access to the system [16]. The forensic firm Mandiant reported in its investigations that many ransomware attacks are enabled due to human errors like using weak passwords and not changing their passwords frequently. Many investigators in the Colonial Pipeline attack reported that one employee likely had a weak password, which could have been infiltrated using standard phishing techniques [41]. Once the password was abducted, the employee did not know about it and continued using it elsewhere, giving the attackers access to multiple access points [21]. The image below explains how poor access management in digital systems is a precursor to cyberattacks, such as ransomware incidents.

CHAPTER 20   CASE STUDIES IN ENERGY

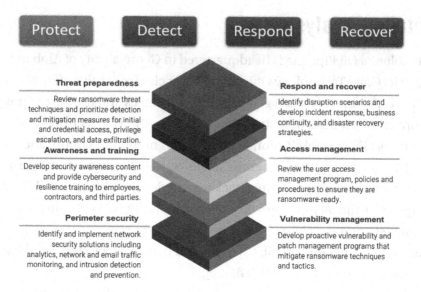

Figure 20-3. *Image Denoting the Main Stages of Building Resilience Against Ransomware [5]. One of the areas for improvement in any interactive digital system is the human interface, where access management fails through poor password management.*

Response and Challenges to Their Implementation

In keeping with the ransomware recovery framework's requirements, the CTO, alongside other IT professionals involved, decided to shut down all operations at the pipeline to prevent the ransomware code from infecting all systems and data. In addition, the company called in the renowned forensic IT firm Mandiant to investigate the technical details [20]. Also, the Federal Bureau of Investigations, Department of Energy, and Department of Homeland Security IT personnel and investigation teams came on board considering the cyberattack's economic and national security implications. However, the company eventually paid DarkSide to get access to a decryption key for several reasons, including the severity of this ransomware attack.

One of the challenges to implementing the ransomware recovery protocols commonly used during such attacks was that the attackers had gained access to several data points and systems. The password used to access Colonial Pipeline's accounting and billing departments had been used several times, but its owner, the company, needed to know if it was compromised or even changed. Also, the mode of operation that DarkSide utilizes made this attack particularly risky, considering the cybercrime group offers ransomware-as-a-service (RaaS). According to Have [6], RaaS operators demand up to 45% of the ransom once their software is utilized for an attack. This mode of cybercrime means they can design and sell ransomware or other forms of malware to threat actors for use in secondary attacks.

Effectiveness of the Strategies Used

Unfortunately, the strategy utilized as part of Colonial Pipeline's ransomware recovery was ineffective for three reasons. First, the password used to access Colonial Pipeline's accounting and billing departments was used several times. At the same time, its owner, the company, did not know it was compromised or even changed it [19]. Second, DarkSide had gained access to several departments and data points, meaning they had control over sensitive and essential data, forcing the company to pay ransom. Third, the decision to bring in Federal law enforcement and security agencies like the DHS and FBI to assist Mandiant and the Department of Energy IT teams in investigating the matter could have been better advised. Such decisions scare many cybercriminals who may compromise the system they had attacked and flee to avoid capture [23]. A typical course of action in ransomware attacks is first attempting negotiations without involving law enforcement. However, the ransomware attack's economic and national security implications would have attracted the interest and participation of Federal agencies regardless.

Case Study 4: Ransomware Attack on the Amsterdam–Rotterdam–Antwerp Refining Hubs

Introduction to the Case

On January 29, 2022, several crude oil refining facilities in Germany, Belgium, and the Netherlands were targeted in a well-designed ransomware attack, causing massive shutdowns and fuel supply problems in many countries [25]. The attack was detected and narrowed down to having attacked at least 17 port cities' software for handling petroleum products in and out of the EU [25]. The ransomware had infiltrated control software to enable automated handling and logistics of petroleum products among oil tankers, port storage facilities, and refineries before offloading to inland transport facilities.

The two companies initially affected by the ransomware before it was spread through cross-company software and hardware operations were Oiltanking GmbH Group and Mabanaft GmbH [25]. These companies stated that they took immediate steps to recover and prevent the malware upon discovering their digital resources were compromised. However, the interconnected nature of port and logistics operations, at both hardware and software levels, meant the malware had likely already spread widely. Further investigations revealed that the Black Cat cybercrime group, which had ties to Revil and DarkSide groups, was responsible for the ransomware attack [26]. The attack caused massive logistics and material handling nightmares by retrogressing automated operations to manual ones at significant ports of entry into the EU.

CHAPTER 20 CASE STUDIES IN ENERGY

Detailed Analysis

The ransomware attack on the Amsterdam-Rotterdam-Antwerp Refining Hubs was designed to use malicious code to take over software used to identify and process oil tankers, pump units, and piping infrastructure used to move petroleum products between logistics, refining, and inland transportation processes. One likely mode of ingress into the VPN-oriented workplace environment was through weak passwords or other human-based weaknesses in digitalized industrial systems [24]. Oil refining and transportation operations at ports have many workers whose credentials are printed on cards they carry and whose access to the digital systems they operate also relies on passwords [42]. Ordinarily, many do not change their passwords as requested by security policies.

Also, the mode of operation at this significant and essential energy installation indicates the presence of initial access brokers (IABs) [27]. These are low-level hackers whose leading trade in the cybercriminal world is gaining access to large industrial or high-security level installations and selling access credentials to larger, more capable cybercrime groups like DarkSide and Black Cat. The demands made after this ransomware attack were higher than the average of 7 million usually demanded, which demonstrated that cybercriminals were becoming bolder due to improved access and capabilities. Finally, forensic investigations into DarkSide attacks on the Colonial Pipeline in the United States and the Amsterdam-Rotterdam-Antwerp Refining Hubs revealed several Russian coding practices and communication habits [32].

Response and Challenges to Their Implementation

After the initial detection of the ransomware attack, the IT teams at the port-based company stations immediately took ransomware recovery actions, like shutting down the software used to process materials at the port and all interconnected hardware [28]. Also, the IT teams isolated the data that was established to have been infected from their backup systems. The first challenge to these responses was the interconnected nature of port digital systems, which compromised significant portions of the entire infrastructure [29]. Second, port-based digital operations involving numerous companies working at a single station comprise thousands of human operators, any of whom were responsible for the system attack through reckless password or access point management.

Effectiveness of Ransomware Recovery Strategies

Unfortunately, the ransomware recovery strategies implemented at the Amsterdam-Rotterdam-Antwerp Refining Hubs had minimal efficiency because of the sheer number of failure points emanating from the thousands of human operators with access to the system. Also, port digital systems are highly interconnected and comprise hundreds of differently secured hardware points with exposed USD ports, meaning shutting one point of the system down may not prevent ransomware code from infecting the entire system [30]. Finally, the cybercrime group involved was competent, implementing other high-level ransomware even on advanced American energy installations like the Colonial Pipeline. Therefore, common ransomware recovery strategies might not work for seasoned and skilled criminals.

References

[1] A. Petrosyan, "Global ransomware victimization rate 2021," *Statista*, Oct. 23, 2023. https://www.statista.com/statistics/204457/businesses-ransomware-attack-rate/

[2] M. Egan, "A Retrospective on 2022 Cyber Incidents in the Wind Energy Sector and Building Future Cyber Resilience," PDF, Boise State University, 2022.

[3] B. Sobczak, "The inside story of the world's most dangerous malware," *E&E News*, Mar. 07, 2019. https://www.eenews.net/articles/the-inside-story-of-the-worlds-most-dangerous-malware/

[4] M. Ahmed and Y.-C. Kim, "Hierarchical Communication Network Architectures for Offshore Wind Power Farms," *Energies*, vol. 7, no. 5, pp. 3420–3437, May 2014, doi: https://doi.org/10.3390/en7053420.

[5] M. Hebert and A. Jowhar, "Improve Ransomware Resilience for Healthcare," *Infotech.com*, 2023. https://www.infotech.com/research/ss/improve-ransomware-resilience-for-healthcare (accessed Jun. 12, 2024).

[6] C. Have, "The Commoditization of Ransomware-as-a-Service," *ISACA*, Feb. 21, 2024. https://www.isaca.org/resources/news-and-trends/newsletters/atisaca/2024/volume-4/the-commoditization-of-ransomware-as-a-service#:~:text=As%20the%20UK (accessed Jun. 12, 2024).

[7] D. C. Smith, "Cybersecurity in the energy sector: are we prepared?," *Journal of Energy & Natural Resources Law*, vol. 39, no. 3, pp. 265–270, Jul. 2021, doi: https://doi.org/10.1080/02646811.2021.1943935.

[8] M. E. McCarty et al., "Cybersecurity Resilience Demonstration for Wind Energy Sites in Co-Simulation Environment," *IEEE Access*, vol. 11, pp. 15297–15313, Jan. 2023, doi: https://doi.org/10.1109/access.2023.3244778.

[9] N. Farrar and Mohd. Hasan Ali, "Cyber-Resilient Converter Control System for Doubly Fed Induction Generator-Based Wind Turbine Generators," *Electronics*, vol. 13, no. 3, pp. 492–492, Jan. 2024, doi: https://doi.org/10.3390/electronics13030492.

[10] M. Borhani, Gurjot Singh Gaba, J. Basaez, Ioannis Avgouleas, and Andrei Gurtov, "A critical analysis of the industrial device scanners' potentials, risks, and preventives," *Journal of industrial information integration*, pp. 100623–100623, May 2024, doi: https://doi.org/10.1016/j.jii.2024.100623.

[11] N. Alhalafi and P. Veeraraghavan, "Cybersecurity Policy Framework in Saudi Arabia: Literature Review," *Frontiers in Computer Science*, vol. 3, Oct. 2021, doi: https://doi.org/10.3389/fcomp.2021.736874.

[12] U. Urooj, B. A. S. Al-rimy, A. Zainal, F. A. Ghaleb, and M. A. Rassam, "Ransomware Detection Using the Dynamic Analysis and Machine Learning: A Survey and Research Directions," *Applied Sciences*, vol. 12, no. 1, p. 172, Dec. 2021, doi: https://doi.org/10.3390/app12010172.

[13] T. Plėta, M. Tvaronavičienė, S. D. Casa, and K. Agafonov, "Cyber-attacks to critical energy infrastructure and management issues: overview of selected cases," *Insights into Regional Development*, vol. 2, no. 3, pp. 703–715, Sep. 2020, doi: https://doi.org/10.9770/ird.2020.2.3(7).

[14] S. Rajeyyagari and A. S. Alotaibi, "A study on cyber-crimes, threats, security and its emerging trends on latest technologies: influence on the Kingdom of Saudi Arabia," *International Journal of Engineering & Technology*, vol. 7, no. 2.3, p. 54, Mar. 2018, doi: https://doi.org/10.14419/ijet.v7i2.3.9969.

[15] T. Alladi, V. Chamola, and S. Zeadally, "Industrial Control Systems: Cyberattack trends and countermeasures," *Computer Communications*, vol. 155, pp. 1–8, Apr. 2020, doi: https://doi.org/10.1016/j.comcom.2020.03.007.

[16] R. Grubbs, J. Stoddard, S. Freeman, and R. Fischer, "Evolution and Trends of Industrial Control System Cyber Incidents since 2017," *Journal of Critical Infrastructure Policy*, vol. 2, no. 2, Dec. 2021, doi: https://doi.org/10.18278/jcip.2.2.4.

[17] M. Gazzan and F. T. Sheldon, "Opportunities for Early Detection and Prediction of Ransomware Attacks against Industrial Control Systems," *Future Internet*, vol. 15, no. 4, p. 144, Apr. 2023, doi: https://doi.org/10.3390/fi15040144.

[18] T. McIntosh, A. S. M. Kayes, Y.-P. P. Chen, A. Ng, and P. Watters, "Ransomware Mitigation in the Modern Era: A Comprehensive Review, Research Challenges, and Future Directions," *ACM Computing Surveys*, vol. 54, no. 9, pp. 1–36, Dec. 2022, doi: https://doi.org/10.1145/3479393.

[19] R. Bold, H. Al-Khateeb, and N. Ersotelos, "Reducing False Negatives in Ransomware Detection: A Critical Evaluation of Machine Learning Algorithms," *Applied Sciences*, vol. 12, no. 24, p. 12941, Dec. 2022, doi: https://doi.org/10.3390/app122412941.

[20] S. Kerner, "Colonial Pipeline Hack explained: Everything You Need to Know," *TechTarget*, Apr. 26, 2022. https://www.techtarget.com/whatis/feature/Colonial-Pipeline-hack-explained-Everything-you-need-to-know

[21] R. A. M. Alsaidi, W. M. S. Yafooz, H. Alolofi, G. A.-M. Taufiq-Hail, A.-H. M. Emara, and A. Abdel-Wahab, "Ransomware Detection using Machine and Deep Learning Approaches," *International Journal of Advanced Computer Science and Applications*, vol. 13, no. 11, 2022, doi: https://doi.org/10.14569/ijacsa.2022.0131112.

[22] D. W. Fernando, N. Komninos, and T. Chen, "A Study on the Evolution of Ransomware Detection Using Machine Learning and Deep Learning Techniques," *IoT*, vol. 1, no. 2, pp. 551–604, Dec. 2020, doi: https://doi.org/10.3390/iot1020030.

[23] A. Corallo, M. Lazoi, and M. Lezzi, "Cybersecurity in the Context of Industry 4.0: a Structured Classification of Critical Assets and Business Impacts," *Computers in Industry*, vol. 114, p. 103165, Jan. 2020, doi: https://doi.org/10.1016/j.compind.2019.103165.

[24] Chalermpong Senarak, "Port cyberattacks from 2011 to 2023: a literature review and discussion of selected cases," *Maritime Economics & Logistics*, vol. 26, no. 1, Dec. 2023, doi: https://doi.org/10.1057/s41278-023-00276-8.

[25] P. Nair, "Cyberattack Cripples European Oil Port Terminals," *www.bankinfosecurity.com*, Feb. 05, 2022. https://www.bankinfosecurity.com/cyberattack-cripples-european-oil-port-terminals-a-18465 (accessed Jun. 12, 2024).

[26] S. J. Root, P. Throckmorton, J. Tacke, J. Benjamin, M. Haney, and R. A. Borrelli, "Cyber hardening of Nuclear Power Plants with real-time nuclear reactor operation, 1. Preliminary operational testing," *Progress in nuclear energy*, vol. 162, pp. 104742–104742, Aug. 2023, doi: https://doi.org/10.1016/j.pnucene.2023.104742.

[27] Ö. Aslan, S. S. Aktuğ, M. Ozkan-Okay, A. A. Yilmaz, and E. Akin, "A Comprehensive Review of Cyber Security Vulnerabilities, Threats, Attacks, and Solutions," *Electronics*, vol. 12, no. 6, p. 1333, Mar. 2023, doi: https://doi.org/10.3390/electronics12061333.

[28] T. R. Reshmi, "Information security breaches due to ransomware attacks—a systematic literature review," *International Journal of Information Management Data Insights*, vol. 1, no. 2, p. 100013, Nov. 2021, doi: https://doi.org/10.1016/j.jjimei.2021.100013.

[29] H. Riggs *et al.*, "Impact, Vulnerabilities, and Mitigation Strategies for Cyber-Secure Critical Infrastructure," *Sensors*, vol. 23, no. 8, p. 4060, Jan. 2023, Available: https://www.mdpi.com/1424-8220/23/8/4060

[30] J.-P. A. Yaacoub, H. N. Noura, O. Salman, and A. Chehab, "Robotics Cyber security: vulnerabilities, attacks, countermeasures, and Recommendations," *International Journal of Information Security*, vol. 21, no. 21, Mar. 2021, doi: https://doi.org/10.1007/s10207-021-00545-8.

[31] IEA, "Cyber resilience—Power Systems in Transition—Analysis," *IEA*, 2020. https://www.iea.org/reports/power-systems-in-transition/cyber-resilience

[32] Allianz, "Cyber attacks on critical infrastructure," *Allianz Commercial*, Jun. 2016. https://commercial.allianz.com/news-and-insights/expert-risk-articles/cyber-attacks-on-critical-infrastructure.html

[33] Resecurity, "Resecurity | Ransomware Attacks against the Energy Sector on the rise—Nuclear and Oil & Gas are Major Targets in 2024," *www.resecurity.com*, Nov. 12, 2023. https://www.resecurity.com/blog/article/ransomware-attacks-against-the-energy-sector-on-the-rise-nuclear-and-oil-gas-are-major-targets-2024#:~:text=%2FALPHV%20(suspected)- (accessed Jun. 13, 2024).

[34] BlackKite, "The 2021 Ransomware Risk Pulse: Energy Sector Ransomware on the Rise Across Critical Infrastructure," Black Kite, 800 Boylston Street, Suite 2905 Boston, MA 02199, 2021.

[35] J. B, L. Tibirna, T. & D. R. T. - TDR, and J. B. TDR Livia Tibirna and Threat & Detection Research Team-, "The Energy sector 2022 cyber threat landscape," *Sekoia.io Blog*, Apr. 05, 2023. https://blog.sekoia.io/the-energy-sector-2022-cyber-threat-landscape/

[36] N. Chen, R. Yu, Y. Chen, and H. Xie, "Hierarchical method for wind turbine prognosis using SCADA data," *IET Renewable Power Generation*, vol. 11, no. 4, pp. 403–410, Mar. 2017, doi: https://doi.org/10.1049/iet-rpg.2016.0247.

[37] D. Silverman, Y.-H. Hu, and M. Hoppa, "A Study on Vulnerabilities and Threats to SCADA Devices," *Journal of The Colloquium for Information Systems Security Education*, vol. 7, no. 1, pp. 8–8, Jul. 2020.

[38] M. IAIANI, A. TUGNOLI, S. BONVICINI, and V. COZZANI, "Analysis of Cybersecurity-related Incidents in the Process Industry," *Reliability Engineering & System Safety*, vol. 209, p. 107485, May 2021, doi: https://doi.org/10.1016/j.ress.2021.107485.

[39] M. Gazzan, A. Alqahtani, and F. T. Sheldon, "Key Factors Influencing the Rise of Current Ransomware Attacks on Industrial Control Systems," *2021 IEEE 11th Annual Computing and Communication Workshop and Conference (CCWC)*, Jan. 2021, doi: https://doi.org/10.1109/ccwc51732.2021.9376179.

[40] J. Lewallen, "Emerging Technologies and Problem Definition uncertainty: the Case of Cybersecurity," *Regulation & Governance*, vol. 15, no. 4, Jul. 2020, doi: https://doi.org/10.1111/rego.12341.

[41] A. M. Y. Koay, R. K. L. Ko, H. Hettema, and K. Radke, "Machine learning in industrial control system (ICS) security: current landscape, opportunities and challenges," *Journal of Intelligent Information Systems*, Oct. 2022, doi: https://doi.org/10.1007/s10844-022-00753-1.

[42] J. Martin and C. Whelan, "Ransomware through the lens of state crime: Conceptualizing ransomware groups as cyber proxies, pirates, and privateers," *State Crime Journal*, vol. 12, no. 1, May 2023, doi: https://doi.org/10.13169/statecrime.12.1.0004.

[43] I. de la Peña Zarzuelo, "Cybersecurity in ports and maritime industry: Reasons for raising awareness on this issue," *Transport Policy*, vol. 100, pp. 1–4, Jan. 2021, doi: https://doi.org/10.1016/j.tranpol.2020.10.001.

[44] J. J. Moreno Escobar, O. Morales Matamoros, R. Tejeida Padilla, I. Lina Reyes, and H. Quintana Espinosa, "A Comprehensive Review on Smart Grids: Challenges and Opportunities," *Sensors*, vol. 21, no. 21, p. 6978, Oct. 2021, doi: https://doi.org/10.3390/s21216978.

[45] D. Fein, "9 Stages of Ransomware & How AI Responds I Darktrace Blog," *darktrace.com*, Dec. 22, 2021. https://darktrace.com/blog/9-stages-of-ransomware-how-ai-responds-at-every-stage#:~:text=Initial%20entry%20%E2%80%93%20the%20first%20stage (accessed Jun. 29, 2024).

CHAPTER 21

Securing Digital Foundations: A Point of View on Cybersecurity in Healthcare

Author:

Fardin Quazi

In the era of digital healthcare transformation, cybersecurity has become a cornerstone for safeguarding patient data and ensuring the integrity of healthcare systems. This chapter introduces the pivotal role of cybersecurity in healthcare administration and management, focusing on protecting digital assets from the myriad of cyber threats prevalent today.

Cybersecurity in healthcare involves more than just protecting data; it's about preserving patients' trust and ensuring the continuous, effective delivery of healthcare services. With the increasing adoption of electronic health records (EHRs), telemedicine, and other digital technologies, the healthcare sector faces unique cybersecurity challenges.

CHAPTER 21 SECURING DIGITAL FOUNDATIONS: A POINT OF VIEW ON CYBERSECURITY IN HEALTHCARE

Key Components of Cybersecurity in Healthcare

1. **Data Protection and Privacy**: Protecting patient information is a legal and ethical obligation for healthcare providers. Not only patient data but also financial records and administrative information require stringent protection. Effective data protection strategies include robust encryption, stringent access controls, and regular monitoring of data access and usage. This ensures compliance with regulations like HIPAA and GDPR while maintaining patient confidentiality.

2. **Risk Management and Compliance**: Healthcare organizations must regularly assess and manage cybersecurity risks. This includes conducting vulnerability assessments, implementing security controls, and ensuring compliance with healthcare-specific regulatory requirements. Regular audits of IT systems, compliance with legal regulations like HIPAA for data protection, and developing policies for data handling and security are important components of risk management and compliance as well. Effective risk management is critical in preventing data breaches and ensuring operational continuity.

3. **Securing Healthcare Networks and Devices**: With the proliferation of IoT devices in healthcare, like wearable health monitors and connected medical devices, securing the network infrastructure is

crucial. This involves deploying firewalls, intrusion detection systems, encryption of emails, securing internal messaging systems, safeguarding electronic transfer of patient data, and securing wireless networks against unauthorized access.

4. **Employee Education and Training:** Healthcare staff often serve as the first line of defense against cyber threats. Regular training on identifying phishing attempts, managing sensitive data, and following best security practices is essential in creating a security-conscious culture.

5. **Incident Response and Recovery:** Developing a comprehensive incident response plan is vital for healthcare organizations. This plan should outline procedures for responding to a cybersecurity incident, minimizing its impact, and recovering from it. It's essential for maintaining patient care and services in the event of a cyberattack. Apart from the standard protocols, the healthcare system should also have a robust communication strategy, including steps for containment, and procedures for data recovery.

In 2020, Magellan Health, a Fortune 500 company specializing in managed healthcare and insurance services, experienced a sophisticated phishing attack. The incident resulted in the exposure of the personal information of employees and patients, including names, treatment information, and employee credentials. The company responded by enhancing its cybersecurity measures, providing identity protection services to affected individuals, and cooperating with law enforcement in the investigation [1]

CHAPTER 21   SECURING DIGITAL FOUNDATIONS: A POINT OF VIEW ON CYBERSECURITY IN HEALTHCARE

Cybersecurity in clinical research as well as healthcare administration and management is crucial for protecting sensitive data and ensuring smooth operations. By focusing on data protection, risk management, secure communications, staff training, and incident management, healthcare entities can mitigate the risk of cyber threats and maintain the trust of their clients and partners.

Threat Landscape in Healthcare

The digitalization of the health sector has brought a new era of interconnected medical devices and systems, known as the "Internet of Medical Things" (IoMT). Apart from the ease of data exchange and optimizing healthcare delivery, IoMT has also brought forth a unique set of cybersecurity challenges. Threat actors are thus able to exploit such vulnerability to gain unauthorized access to sensitive patient data or even to disrupt healthcare services through cyberattacks. Apart from this, there are hosts of new-age sophisticated threats to cybersecurity, targeting electronic health records (EHRs), telehealth services, and other sensitive health data and critical infrastructure of the healthcare systems.

With the rise of artificial intelligence and machine learning in the healthcare environment, new threat landscapes continue to emerge. Advanced attacks can even manipulate the algorithms leading to incorrect diagnoses or treatment planning. We analyze a few instances of the threat landscape in healthcare.

1. **Phishing Attacks:** These are the most common cybersecurity threats in healthcare. Cybercriminal sends innocuous-looking emails to the employees of a healthcare organization with links that trick the user into disclosing sensitive information. Often, such emails use a well-known medical disturbance to incentivize the clicking of links.

2. **Ransomware Attacks:** Another growing threat among healthcare providers is ransomware attacks. In this kind of attack, the malware is sneaked into a network to infect and further encrypt all the sensitive data until a ransom is paid for its decryption. Such malware is commonly injected into the system through a phishing attack.

3. **Insider Threats:** Insider threats can emanate within a healthcare organization. The threats can be malicious, where the insiders would willingly cause harm, or where the employees unintentionally make mistakes that may lead to security breaches.

4. **Medical Device Hacking:** The emergence of the Internet of Medical Things (IoMT) has come with new cybersecurity challenges. Vulnerabilities on such devices, if targeted by threat actors, might lead to unauthorized access to critical patient data or disrupt health services.

5. **Insecure IoT Devices:** A large number of IoT devices across healthcare facilities expose the entire infrastructure susceptible to cyberattacks. Poorly secured IoT gadgets in use could be targeted for gaining access to health systems in such an environment.

6. **Lack of Security Awareness:** Ignorance about policies and regulations in place and how they influence the handling of patients' data, how should the information be processed, transmitted, and stored in a Smart healthcare setup exponentially increases the threat landscape, making it a soft target for the attackers.

CHAPTER 21   SECURING DIGITAL FOUNDATIONS: A POINT OF VIEW ON CYBERSECURITY
            IN HEALTHCARE

Let's view a couple of case studies that underscore the importance of robust cybersecurity measures in the healthcare sector. They provide valuable insights into the types of threats healthcare organizations face and the potential consequences of these threats.

1. **UVM Health Network Ransomware Attack:** In October 2020, the University of Vermont (UVM) Health Network, a six-hospital healthcare organization serving over 1 million patients, fell victim to a ransomware attack. The attack, which stemmed from an employee error, led to significant disruptions across the organization's infrastructure, shutting down critical technology and delaying patient care [2].

2. **Shields Healthcare Cyberattack:** In March 2022, Shields, a healthcare service provider, suffered a cyberattack that exposed the data of 2 million patients. The attack impacted almost 60 affiliated healthcare facilities, including well-known hospitals, medical centers, and clinics [3].

Security Principles in Healthcare

As the digital landscape continues to rapidly advance, health organizations are increasingly adopting advanced technology. From electronic health records to telemedicine and real-time data monitoring to diagnostic tools, all have been brought to the forefront of increasing adoption and use. This surge in digitalization highlights the critical need for robust cybersecurity measures that address the needs of a highly sensitive and regulated healthcare system. The nucleus of these security efforts is defined by the fundamental principles of confidentiality, integrity, and availability — known as the CIA triad — which ensure that patient data is private, precise, and readily accessible, especially critical in life-saving scenarios.

CHAPTER 21   SECURING DIGITAL FOUNDATIONS: A POINT OF VIEW ON CYBERSECURITY IN HEALTHCARE

The principles of healthcare security strategy are further realized through the application of the defense-in-depth and least-privilege. They minimize the risk of data breaches by incorporating multilayer security mechanisms to safeguard the data while restraining user access to essential functions. The essential part of this security setup is the encryption of every data bit, ensuring that the patient's information remains confidential in storage and during transmission. This ensures the protection of data privacy, thereby establishing trust between patients and digital health interactions.

Continuous monitoring and proactive vulnerability management are essential to timely identification and resolution, making healthcare systems more resilient than reactive to cyber threats. Regular security audits, and ongoing staff training, help imbibe a culture of security awareness and for the maintenance of a robust defense against cyber disruptions.

Implementation of these comprehensive cybersecurity protocols is not only a technical compulsion but also a cornerstone requirement in modern healthcare delivery, ensuring safety in handling patient data and uninterrupted medical services. This commitment to cybersecurity principles of healthcare infrastructure is indispensable to secure and trustworthy patient-centric digital healthcare systems.

Ransomware in Digital Healthcare Landscape

The role of ransomware in a dynamic environment of digital healthcare is not just paramount but of huge significance in the current surge of cyber threats across the globe. Ransomware is a very sophisticated and malicious malware attack, which is aggressive and poses severe risks to an individual and an organization, particularly in the area of health. The acuteness and sensitivity of healthcare organizations dealing with sensitive patients' data rest on the integrity of the medical systems and safeguarding of the PHI.

CHAPTER 21 SECURING DIGITAL FOUNDATIONS: A POINT OF VIEW ON CYBERSECURITY IN HEALTHCARE

Ransomware is a type of malicious software designed to block access to a computer system or important data on a computer or over a network through encryption, under a ransom plea for the restoration of access after making a digital currency payment. Such operations can disrupt healthcare services, compromise patient data privacy, and cause heavy financial loss.

The healthcare industry's increasing dependency on digital technology makes it vulnerable to ransomware attacks. With the growing adoption of electronic health records (EHRs), telemedicine, and other digital health services, the sheer volume of storage and exchange of sensitive data has grown enormously. This data, which is the backbone of delivering quality patient care and reliable medical procedures, becomes a nefarious tool for cybercriminals.

Understanding the Entry Points of Ransomware

Ransomware often infiltrates systems through numerous channels, with phishing attacks being the most common. Typically, this would involve sending emails that would look authentic and hence mislead the target to follow a link or download an attachment that has a malware infection. Once the malware is inside the system, it can quickly spread across the network, encrypting data and locking out users.

Other entry points include unpatched software vulnerabilities, weak password hacks by brute forcing, and insecure remote desktop protocols. If these vulnerabilities are not enough, healthcare organizations frequently face challenges in regular updating and patching of their systems, making them more likely to fall prey to such attacks.

CHAPTER 21   SECURING DIGITAL FOUNDATIONS: A POINT OF VIEW ON CYBERSECURITY IN HEALTHCARE

Impact on Healthcare: The Consequences of Ransomware in Digital Healthcare

In digital healthcare, ransomware can go beyond a mere data breach and threaten the core of patient care and health safety. But for healthcare providers, addressing such a cyberattack means even more daunting challenges than the loss, compromise, or control of data; it means a real and substantial risk to the essence and fundamentals of medical practice—means and resources for giving safe and efficient care to the patient.

Ransomware attacks can cripple the entire service delivery of a healthcare ecosystem. The results of cutting access to PHIs, important healthcare systems, and medical devices are immediate and devastating. The most important factor is personalized and accurate medical attention based on the patient's health records. Without such records, healthcare providers will work in an absolute vacuum, resulting in delays and potential misdiagnosis in treatment. It endangers the life of a person, especially in critical care where every second counts.

The impact of ransomware on healthcare goes far beyond the immediate effects it has on patients' care. The financial consequences are manifold, reaching far from ransom payments often demanded in cryptocurrency. Furthermore, there is no guarantee that even if the ransom is paid, access to the encryption of data will be restored. After paying off the ransom, a lot of financial resources will be required for the restoration of systems and data return, if possible. The process is not only costly but time-consuming, often requiring specialized expertise.

These inabilities represent another physical and substantial financial strain. When healthcare systems are offline, the regular flow of patient care is interrupted. This may result in disruptions, cancellations of appointments, delays in providing elective surgeries, and reduced capacity to accept new patients, which are all directly proportional to revenue losses. The financial implications of such cyberattacks may be devastating to the smaller healthcare facilities, threatening their very establishment. Additionally, there is an impact of long-term reputational damage. Trust is a foundational element in healthcare; patients leave their most delicate information to healthcare providers, believing that it is going to be safeguarded. Ransomware breaches smash that trust and drive patients potentially out of the care of the clinic. Rebuilding this trust is a long, hard process that requires transparent communication and ongoing demonstration of improvements in cybersecurity measures [4].

Moreover, ransomware attacks in healthcare settings can have legal ramifications. The health provider is bound by a regulated framework, such as the HIPAA (Health Insurance Portability and Accountability Act) in the United States, which mandates the protection of patient information. Ransomware breaches can lead to incompliance with these regulations and can be consequential in hefty fines and ensuing legal battles. The widespread impact of ransomware demands a proactive strategy in digital healthcare administration. This involves investing in advanced cybersecurity measures, providing regular training for healthcare staff on cyber hygiene, maintaining systematic backups, and implementing robust incident response plans. Strengthening defenses also requires collaboration with cybersecurity and regulatory compliance experts to safeguard against these evolving threats.

Healthcare providers must engage in sector-wide sharing of information and best practices regarding ransomware threats. The healthcare sector can reinforce its defenses by understanding and analyzing the evolving nature of these cyberattacks through collaborative efforts. Ransomware in digital health administration is a multidimensional threat with severe and boundless implications. It influences the financial stability of healthcare institutions but most essentially influences the quality and safety of care for patients. Addressing this demon of ransomware calls for an all-inclusive multilateral approach in prevention, preparedness, and response to make sure that the healthcare sector is resilient to digital threats.

Ransomware Prevention in a Digital Healthcare System

This section of the chapter focuses on the implications and the importance of preventive measures against ransomware attacks on health institutions and provider organizations. In 2023, ransomware compromised at least 141 hospitals across 46 systems, disrupting operations by forcing emergency departments to redirect services, delaying treatments, and hindering access to medical records [5]. The economic toll is substantial, with an estimated $77.5 billion lost in downtime since 2016, excluding costs associated with litigation and the erosion of public trust [6].

With advancements in digital technologies in healthcare creating a network of interconnected systems and devices, the risk of ransomware increases simultaneously. Ransomware attacks on healthcare systems are not only damaging the critical healthcare infrastructure but also expose sensitive patient data, resulting in far more serious consequences. The criticality of data and system downtime necessitates a robust ransomware prevention strategy and operational plan. We discuss a few preventive measures here:

CHAPTER 21 SECURING DIGITAL FOUNDATIONS: A POINT OF VIEW ON CYBERSECURITY IN HEALTHCARE

Figure 21-1. Ransomware Preventive Measures in Healthcare (Image Idea: Freepik.com)

1. **Data Backup:** A simple yet effective practice, to avoid disruption, is data backup. Healthcare organizations should ensure that frequently updated backups are maintained either independently or in cloud services so that in the event of compromise, the data can be restored promptly without falling prey to ransom demands. Periodic testing along with updates is vital to ensure that data restoration and executing BAU measures are efficient and effective.

2. **Employee Training:** Ransomware breaches usually come in via phishing emails. Training the staff as well as 3rd part vendors, patients, and other stakeholders is the most important aspect of threat

awareness and prevention. Knowledge sharing with the staff through hands-on exercise and keeping them abreast of the security measures and protocols can reduce the system and data vulnerability to a great extent.

3. **System Maintenance:** Healthcare systems should ensure that software, applications, and medical devices containing current or latest security patches are updated. This proactive cyber threat management will help in curtailing the damage likely to be caused by ransomware. Rapid patch management should be prioritized, especially for the systems accessible via the Internet, e.g., patient portals or telemedicine service systems.

4. **Advanced Threat Detection:** It enables the system to be predictive against potential ransomware activities. It strengthens healthcare organizations to analyze early warning signs and undertake timely action that can be taken before the spreading of the malware.

5. **Incident Response Planning:** Properly defined incident response planning is the other critical factor. It is supposed to clearly outline the steps that are meant to be followed in case of an attack to minimize damage and even operational downtime.

These proactive measures ensure that the systems are protected from any ransomware attacks, thus ensuring safe digital transformation processes are well-defined while mitigating any occurrences of cyberattack. A multitier defensive approach ensures the safety of sensitive operations and data.

CHAPTER 21 SECURING DIGITAL FOUNDATIONS: A POINT OF VIEW ON CYBERSECURITY IN HEALTHCARE

Figure 21-2. *UVM Health Network Cyberattack (Image Source: https://www.wcax.com/2020/10/29/uvm-health-network-a-victim-of-cyberattack/)*

In October 2020, UVM Health Network, a six-hospital healthcare organization serving over 1 million patients throughout Vermont and upstate New York, discovered that its systems had been compromised by a ransomware attack [7]. The attack led to major disruptions across the organization's infrastructure, shutting down critical technology and delaying patient care.

The attack stemmed from an employee error. An employee took their work laptop on vacation and used it to check their personal emails. One of these emails was a phishing scam, and by opening the email, the employee unknowingly allowed cybercriminals to launch malware on their work laptop.

When the employee returned to work and connected their laptop to the UVM Health Network's systems, the cyber criminals utilized that malware to target the entire organization1. On October 28, the cybercriminals officially launched their attack on UVM Health Network, spreading malware across the organization's technology.

CHAPTER 21   SECURING DIGITAL FOUNDATIONS: A POINT OF VIEW ON CYBERSECURITY IN HEALTHCARE

UVM Health Network immediately went offline to protect its sensitive records. However, the attack resulted in significant recovery costs and reputational damages for UVM Health Network. This case study emphasizes the importance of employee education and robust cybersecurity measures in preventing ransomware attacks.

Ransomware Detection and Response: Early Detection Techniques in Healthcare Administration and Management

Healthcare is at the forefront of tangible service disciplines that face numerous challenges in this modern era of digital transformation, where cybersecurity is identified as a key concern. The increasing spate of cyberattacks in the form of ransomware makes early detection of these threats not only a technical necessity but also an aspect of patient care and data security.

The sophistication of this ransomware is evolving rapidly. The attackers are inconsiderate of the data vitality by blocking access to critical patient records, interference with the medical processes, and exposure or tampering with private and sensitive data. These attacks could be appalling not only for operational aspects of the healthcare facility but also for patient safety. Early detection of ransomware in healthcare settings involves a multifaceted approach, integrating both technological and human elements. We will discuss a few of such techniques and approaches here:

1. **Advanced Monitoring Systems in Healthcare IT Infrastructure:** Early recognition of ransomware threats to health administration and management through advanced monitoring systems is paramount. It is a specially designed system meant

for the continuous monitoring of all healthcare IT networks containing sensitive data regarding patients. These systems will detect anomalies differing from the normal network behavior, including unexpected encryption of EHR (electronic health record) of patient records, spikes of transfer data rates, or unauthorized attempts to access healthcare databases.

The healthcare sector requires foolproof defense levels with enormous repositories of confidential patient information. Thus, these monitoring tools not only become technological but also methodical safeguards in the wider view of the healthcare cybersecurity strategy. They also act as a flagging system for possible breaches in real-time, something highly needed for early warning in healthcare. Any delay in the detection of a breach may mean not only data loss but also a danger to patient care and privacy.

Successful implementation involves strategic alignment that must include compliance standards in healthcare, such as the Health Insurance Portability and Accountability Act (HIPAA) and the General Data Protection Regulation (GDPR). These tools are configured to monitor network traffic for signs of ransomware deployment, such as the rapid encryption of files, which is a hallmark of ransomware attacks. They look for unusual patterns of access that could flag the attackers attempting to navigate the network to locate sensitive data.

The effectiveness of these monitoring systems in healthcare settings has been documented in several instances. For example, an advanced monitoring system can help in the early detection of a ransomware attack on a large hospital network, giving ample time to the IT staff to isolate the affected systems and hence prevent the wide compromise of data. These systems not only provide real-time alerts to potential threats, but they also ensure adherence to some of the most crucial regulations safeguarding healthcare data and, eventually, patient's health information and continuity of care."

2. **Regular System Audits and Assessments in Healthcare IT Management [8]:** Conducting regular system audits forms a critical defense line against ransomware threats. These systematic assessments of the healthcare IT systems, processes, and frameworks are crucial in the detection of potential susceptibilities that could be exploited by malicious software like ransomware. The whole process of the audit considers a deep analysis of the IT environment of the healthcare organization. It thoroughly examines the software's different versions to identify the ones that are outdated and liable to security breaches. The outdated software often lacks some of the latest security updates and, hence, is susceptible to these ransomware attacks. It becomes one of the most important preventive controls to ensure that all software is updated and fully patched. This facet of IT management

syncs closely with the regulatory requirements in healthcare, such as those mandated by HIPAA, requiring stringent adherence to PHI safeguards. Furthermore, these audits include an assessment of the organization's overall security measures. This includes looking at the configuration of the firewall, assessing the control of accesses, and reviewing the effectiveness of the information security policies currently in place. The aim is to establish a complete perspective on the preparedness of the healthcare institution from a cybersecurity aspect and pin down the areas needing improvements. There is enough documentary evidence that advocates the effectiveness of routine systems audits in bringing down the many cases of ransomware attacks within the healthcare environment. Numerous studies have been carried out by healthcare IT security establishments, which highlight the importance of routine audits and how they are able to identify the weaknesses in the network of a healthcare facility. This proactive approach proves that possible data breaches and failure in operation are preventable, allowing the institutions to beef up their security measures even before the attack by ransomware occurs. This ensures continuous vigilance of the cybersecurity landscape and can protect the integrity of patient's data and the seamless operations of healthcare services.

3. **Incident Response Protocol in Healthcare Organizations [9]:** Developing a well-framed incident response protocol framed to ransomware threats becomes an important aspect of

administration in healthcare. A successful and prompt reaction to ransomware attacks is not a technological necessity but an integral part of patient care that cannot be ignored. The response protocol should ensure that the infected systems are isolated immediately to avoid any further spread of the ransomware. However, interfaced healthcare information systems within the health facility may compromise patient data, hence for that matter, critical healthcare operations throughout the health facility may also be compromised. A well-documented protocol should guide on what steps should be taken and by whom, in particular, in disconnection from the network of the affected systems.

Next, the plan must include procedures for securing unaffected resources. This will involve ensuring that there is backup system availability and other critical infrastructures required for the delivery of healthcare service. The unaffected areas must stay in operation and remain secure from the lapse in patient care during a ransomware event. Communication is equally important. There should be a guideline on communication to various stakeholders: patients on potential data breaches and service disruptions; regulatory bodies with reports according to laws like HIPAA; and also law enforcement on the investigations and remedial actions. Effective communication is key to maintaining trust and transparency during and after a cybersecurity incident.

CHAPTER 21 SECURING DIGITAL FOUNDATIONS: A POINT OF VIEW ON CYBERSECURITY IN HEALTHCARE

For example, let us take a sample of a strong incident response protocol concerning an incident of a ransomware attack on the hospital management system. This incident once reported, thus marking quick actions by the hospital in rolling out its response plan through the isolation of affected systems and notifying the stakeholders of the challenge to reduce impacts from the attack and avail them with quicker restoration of services. This ensures the protection of patient's data, the preservation of trust, and the continuity of essential healthcare services during the crisis involving information and cyberspace through swift, coherent, and effective work of response operations.

4. **Employee Training and Awareness Programs in Healthcare:** The human element in healthcare cybersecurity has to be managed efficiently. A healthcare organization should, therefore, focus its attention and resources on employee training and awareness programs, which are critical in the defense against ransomware attacks. A major cause of such attacks is often the outcome of human errors; hence, training and educating the whole staff is indispensable in the healthcare field. The training programs should be conducted regularly to keep workers updated with the new ransomware tactics. These include ensuring the proper guidance and training of all the staff teaching fraternity to be in a position to identify phishing emails, suspicious links, and weird behavior in the system, which in

most cases are common vectors of ransomware. Employees are also to be trained in safe password practices and risks associated with installing unauthorized software.

It also forms an important part of these training programs: that employees are made capable enough to report such activities. Taking proactive steps may lead to timely detection, which would be able to improve a cybersecurity culture among the employees. Making cybersecurity a shared responsibility will help to improve the total defense for healthcare providers against threats to their medical data, which could prove devastating if exposed.

Data and metrics play a pivotal role in underscoring the importance of these training programs. For instance, a study by the Ponemon Institute found that 88% of ransomware attacks[10] in healthcare organizations were attributed to human error or behavior. Furthermore, a report by health IT security indicated that organizations with regular cybersecurity training reduced the risk of a successful ransomware attack by up to 70% [11]. Moreover, real-life examples demonstrate the effectiveness of these programs. In one case, a large healthcare provider implemented a comprehensive staff training program, resulting in a significant decrease in phishing email click rates within a year, significantly reducing the risk of ransomware infections.

Employee training and awareness programs in healthcare are not limited to knowledge dissipation; they intend to build a vigilant and responsive workforce that can serve as the first line of defense against such ransomware threats. This is very crucial, not only in the protection of sensitive patient data but also for the continuity of critical healthcare services.

A few other techniques which can be implemented for early detection of Ransomware attacks are:

5. **Regular Software Updates and Patch Management:** Keeping all software and systems up to date with the latest patches is a critical defense against ransomware. Many ransomware attacks exploit known vulnerabilities in outdated software, so maintaining up-to-date systems can significantly reduce the risk.

6. **Network Segmentation:** Implementing network segmentation can limit the spread of ransomware if an attack occurs. By dividing the network into separate segments, healthcare administrators can isolate critical systems and data, reducing the potential impact of an attack.

The early detection of ransomware in a healthcare setup is imperative. It requires a holistic approach that combines advanced technology, regular risk assessments, employee training, robust data backup strategies, effective incident response, collaboration with cybersecurity experts, diligent software maintenance, and strategic network management. By prioritizing these measures, healthcare organizations can not only detect ransomware threats early but also mitigate their impact, ensuring the safety and privacy of patient data and the continuity of critical healthcare services.

Analyzing the Data Trends in Healthcare Ransomware Attacks

The rise in ransomware attacks on the healthcare sector is a disturbing trend that poses significant threats not only to the security of sensitive patient data but also to the provision of critical healthcare services. Healthcare ransomware attacks have increased manifold over the last many years, illustrating the growing scale and frequency of these cyber threats and underscoring the severe financial implications in addition to the potential risks to patient safety and privacy.

This escalation can be attributed to several factors. The healthcare sector possesses a wealth of sensitive patient data, making it a lucrative target for cybercriminals. Personal health information (PHI) holds significant value on the dark web, often more so than credit card information, due to its comprehensive nature, which includes insurance details, personal identification data, and medical histories.

Moreover, the transition to digital healthcare systems, while beneficial in terms of efficiency and data accessibility, has also increased the vulnerability of healthcare institutions to cyberattacks. The rapid implementation of electronic health records (EHRs), telemedicine, and other digital health technologies has often outpaced the development of corresponding cybersecurity measures. Many healthcare providers find themselves with outdated security protocols, unpatched software vulnerabilities, and inadequate employee training on cybersecurity, all of which create potential entry points for ransomware attacks.

CHAPTER 21 SECURING DIGITAL FOUNDATIONS: A POINT OF VIEW ON CYBERSECURITY IN HEALTHCARE

The impact of these attacks extends beyond financial losses. When a ransomware attack hits a healthcare institution, it can result in the encryption of crucial patient records and hospital operational systems. This disruption can lead to the cancellation of medical procedures, diversion of emergency patients, and, in severe cases, could potentially result in endangering patient lives. For example, in the event of an attack, access to critical patient information can be delayed or made entirely unavailable, which may hinder the ability of healthcare professionals to make informed decisions and provide timely care.

Healthcare institutions have started recognizing the importance of robust cybersecurity measures as a part of their operational integrity. Investments are being made in advanced security solutions such as AI-driven threat detection systems, regular cybersecurity audits, and employee training programs focused on identifying and preventing potential cyber threats. In addition to technological solutions, there is a growing emphasis on developing comprehensive incident response strategies to quickly and effectively respond to ransomware attacks, minimizing the impact on patient care and operational continuity.

The increasing trend of ransomware attacks in the healthcare sector, with its consequent financial and operational repercussions, serves as a call to action for enhanced cybersecurity measures. Healthcare institutions must prioritize the implementation of robust cyber defense mechanisms, regular system updates, staff training, and incident response planning to safeguard against the escalating threat of ransomware attacks. As digital technology continues to permeate the healthcare industry, a proactive approach to cybersecurity is not just a matter of protecting data but a critical component of ensuring patient safety and maintaining trust in healthcare systems.

CHAPTER 21   SECURING DIGITAL FOUNDATIONS: A POINT OF VIEW ON CYBERSECURITY IN HEALTHCARE

Incident Response: The Lifeline of Digital Healthcare

The criticality of incident response (IR) in digital healthcare systems is not limited to the IT infrastructure but also to save the lifeline of healthcare services: patient data and care delivery. The digital transformation in healthcare has resulted in increased adoption of electronic health records (EHRs), telemedicine, and mobile health applications, making incident response an integral part of any healthcare administration setup. The heavy reliance on digital technology by organizations in the health sector is making them a target for cyber threats. These threats range from data breaches to ransomware attacks, with the potential to hold both patient data and important services at risk unless firms are prepared with strong incident response plans.

The focus of a robust incident response workflow and management should aim to swiftly and effectively reduce the threat incidents. This includes identification of potential risks, containment of the impact, eradication of the threat from the system, and restoration of normal operation. As a proactive measure, it is necessary to conduct an in-depth analysis to understand the cause and avoid future occurrences in the wake of an incident.

Digital transformation has brought new tools and methodologies to incident response. Advanced analytics, artificial intelligence, and machine learning are some of the technologies that will be leveraged to predict and identify threats, respond automatically, and work toward the betterment of healthcare organizational security systems. Effective incident response in healthcare means quickly identifying threats, assessing their impact, and recovering rapidly to protect patient information. This proactive approach is essential because data breaches can cause not only financial loss but also harm to a healthcare organization's reputation and patient safety. Hence, integrating advanced cybersecurity measures, regular staff training, and updated response protocols should be implemented to ensure continued trust and compliance in the current digitized healthcare environment.

The February 2024 major cyberattack on Change Healthcare presents an apt example to illustrate the need for an explicit incident response mechanism [12]. The attack hampered Change Healthcare's ability to administer customer payments and insurance claims causing tremendous disruption. This cybercrime episode underscores the need for strong incident response strategies in healthcare. It demonstrated that healthcare organizations need to have protocols in place to quickly identify potential threats, contain the impact, eradicate the threat from the system, and restore normal operations. The incident also underlines the role of employee training in cybersecurity. Employees of healthcare organizations should be educated and made aware of the possible cyber threats and their role in the event of such an occurrence to function as a cohesive safeguard of organizational safety.

CHAPTER 21   SECURING DIGITAL FOUNDATIONS: A POINT OF VIEW ON CYBERSECURITY IN HEALTHCARE

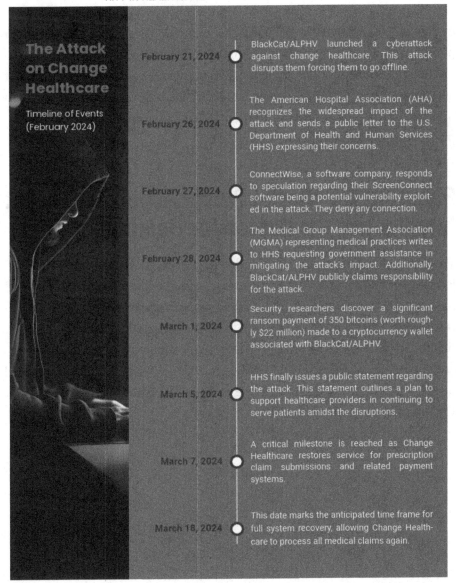

Figure 21-3. *Timeline of Cyberattack on Change Healthcare in February 2024 (Image Source: caplinehealthcaremanagement.com)*

CHAPTER 21 SECURING DIGITAL FOUNDATIONS: A POINT OF VIEW ON CYBERSECURITY IN HEALTHCARE

In conclusion, as healthcare continues toward the rapid advancement in digital technology over the next decade, incident response will be a vital part of overall security and management. Incident response is not just about a reactive approach in responding to threats but creating a culture of awareness and preparation for security and upholding trust in a healthcare organization. These are the fundamental aspects of healthcare delivery prioritizing patient data protection.

Securing the Digital Pulse: Advanced Threat Intelligence Strategies for Healthcare

Threat intelligence is a key enabler driving the digital transformation journey for healthcare. The worrisome trends in cyberattacks in the recent past had a crippling impact on healthcare organizations. Worldwide ransomware attacks inflated by 74% in 2023 compared to 2022. The healthcare sector was not spared either, as it came close to doubling the count of ransomware attacks, from 214 in 2022 to 389 in 2023. In the United States, attacks against the healthcare sector alone rose 128%, with 258 victims in 2023 compared to 113 in 2022 [13].

CHAPTER 21   SECURING DIGITAL FOUNDATIONS: A POINT OF VIEW ON CYBERSECURITY IN HEALTHCARE

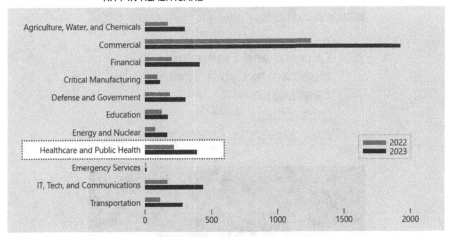

Figure 21-4. Comparison of Global Ransomware Attacks by Sector (2022 vs. 2023) (Image Source: Ransomware_Attacks_Surge_in_2023.pdf (dni.gov))

These are serious potential impacts. In the event of a ransomware attack, organization operation is disturbed, which can paralyze patient care. The resulting financial implications include the cost of ransom, remediation cost, brand damage, etc. In the fast-evolving landscape of healthcare administration, organizations take innovative and advanced technological measures to develop threat intelligence for cyber defense. **Artificial intelligence (AI)** and **machine learning (ML)** techniques are being used to forecast and eliminate cyber threats. They help in identifying patterns and anomalies in the environment that may signal a cyber threat, correspondingly allowing proactive defense.

Another innovative measure is the **threat intelligence platforms (TIPs)**. The TIPs analyze data collected from several sources and transform it into really useful information for the security teams. Healthcare organizations also focus on **securing endpoints and networks, encrypting sensitive data, and training personnel** to recognize and mitigate cyber risks. Regulatory frameworks and standards mandate certain cybersecurity measures to protect patient data.

CHAPTER 21 SECURING DIGITAL FOUNDATIONS: A POINT OF VIEW ON CYBERSECURITY IN HEALTHCARE

Figure 21-5. *Health-ISAC's 2023 Annual Cyber Threat Report—Cover Page (Image Source: 2023 Health Cybersecurity Annual Threat Report - Health-ISAC - Health Information Sharing and Analysis Center)*

The Health-ISAC's 2023 annual cyber threat report [14] provides a comprehensive overview of the current and emerging cyber threats to healthcare organizations, which is fundamental to developing a robust threat intelligence framework. Leveraging the analysts of Booz Allen Hamilton, Health-ISAC created a report on the most diverse and experienced perspective.

CHAPTER 21 SECURING DIGITAL FOUNDATIONS: A POINT OF VIEW ON CYBERSECURITY IN HEALTHCARE

The report discusses major cyber threats for healthcare organizations with an intent to give advice, particularly to senior leaders and practitioners of the health sector in shaping their strategic cybersecurity budget and investment decisions. It reviews and analyzes the current cybercriminal, geopolitical, and nation-state threats that face healthcare organizations, as well as those threats on the horizon, like product abuse and synthetic accounts.

Among the findings, one of the important points is the increase in connected medical devices, which means an increase in the hospital attack surface. While newer medical device designs have improved cybersecurity controls, organizations must prepare for and defend across a broad spectrum of technology while also dealing with software end-of-life issues in legacy medical devices.

The report also highlights the results of a survey conducted in November 2022, where executives across Health-ISAC, CHIME, and the Health Sector Coordinating Council rank ordered the Top Five "greatest cybersecurity concerns" facing their organizations for both 2022 and 2023. This survey engaged cyber (e.g., CISO) and noncyber executives (e.g., CFO) and subsectors such as Providers, Pharma, Payers, Medical Device Manufacturers, and Health IT. It also engaged healthcare organizations with several sizes and varying IT/IS budgets.

CHAPTER 21   SECURING DIGITAL FOUNDATIONS: A POINT OF VIEW ON CYBERSECURITY IN HEALTHCARE

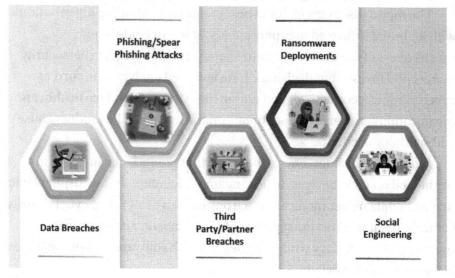

Figure 21-6. *Top Five "Greatest Cybersecurity Concerns" (Image Idea: Freepik.com & 2023 Health Cybersecurity Annual Threat Report - Health-ISAC - Health Information Sharing and Analysis Center)*

This Health-ISAC annual cyber threat report provides a detailed explanation of current, growing, and new potential cyber threats to healthcare organizations, while they are preparing and improving their existing cybersecurity resilience measures.

Healthcare remains one such area where the scope of threat intelligence is vast. It involves the identification of systemic risks in the health sector concerning key suppliers and sector concentration risks and discerning lessons learned to update incident response plans.

Looking forward, future trends for cybersecurity point to the fact that healthcare organizations must be proactive and prepared for a surge in cyberattacks especially in the majority of incidents where network access was breached. The initial access was a result of an external remote service, necessitating an increase in the need for endpoint security. Thus, healthcare threat intelligence strategies have to evolve as the new wave of cyberattacks takes root. The pathway to comprehensive healthcare has to be characterized by eternal vigilance, agility, and an understanding of the threat landscape of historical and modern cyberspace at a very intimate level.

Future of Digital Security in Healthcare

As the healthcare industry undergoes a significant digital transformation, the integration of advanced technologies has brought forth both unprecedented capabilities and new vulnerabilities. This evolution necessitates a robust approach to digital security, focusing not only on combating current threats but also on anticipating and mitigating future vulnerabilities.

The Digital Transformation Landscape in Healthcare

Consumer-Centric Approach

Digital capabilities are becoming essential tools for healthcare systems eager to transform their relationships with consumers. Healthcare is becoming more consumer-focused, and organizations are rethinking what digital experiences can mean for the patient experience. This shift requires designing processes from the consumer's perspective, building trust, loyalty, and introducing innovative forms of digital care delivery. However, this increased digital interaction also expands the attack vectors, making it more imperative to protect confidential patient data with secure measures.

CHAPTER 21   SECURING DIGITAL FOUNDATIONS: A POINT OF VIEW ON CYBERSECURITY IN HEALTHCARE

Interim Milestones and Value Measurement

A digital transformation journey is about setting interim milestones and proving the value of digital initiatives. Health systems frequently assess these milestones as part of their effort to measure ROI and to evolve their security strategies in response to the ever-evolving nature of threat risks. This is imperative for the health system to adopt these strategies of continuous improvement.

Talent and Data Challenges

The success of digital transformation heavily relies on the quality and availability of talent. Healthcare organizations prioritize the recruitment and development of skilled professionals whose roles are focused on analyzing the complexities of digital technologies and cybersecurity. Eliminating data interoperability challenges and defining right key performance indicators (KPIs) will ensure in upholding data security and governance.

As the healthcare systems move to be consumer-centric, the importance of a robust digital security setup has become more critical. Implementing digital solutions to enrich patient experiences is accompanied by increased exposure to cyber threats. This has in turn led to the implementation of high-tech security measures for the safe data storage of their patients and the stability of their online systems. As they navigate through various stages of their digital transformation journey, continuously evaluating and fortifying their cybersecurity frameworks are essential to safeguard against evolving threats and maintain consumer trust.

CHAPTER 21   SECURING DIGITAL FOUNDATIONS: A POINT OF VIEW ON CYBERSECURITY IN HEALTHCARE

Future Trends in Healthcare Cybersecurity

As the healthcare sector moves to advanced digital technologies such as electronic health records (EHRs), telehealth and telemedicine services, and AI-driven applications and tools, it also makes it vulnerable to increased cyberattacks. Take the examples of devices leveraging the IoMT (Internet of Medical Things) with multiple users, access points, and frequent data updates, making them susceptible to cyber threats, necessitating a robust and encompassing digital security framework. Let us analyze a few emerging trends molding the next-gen digital security safeguards in a healthcare organization.

Predictive Analytics and Machine Learning

Timeliness and real-time response to threats are not sufficient to stop the highly sophisticated cyberattacks. The use of predictive analytics built on machine learning not only helps prevent systems from falling prey to such threats but also creates patterns of vulnerable scenarios, which helps healthcare organizations to design a risk-mitigation strategy to shield their digital setup from such attacks. These technologies can also analyze huge amounts of data and scenarios of past security breaches to create a robust and proven response strategy.

Future Plans and Preemptive Strategies

Proactive Cybersecurity Frameworks
Advanced healthcare systems handling sensitive and highly personal patient data require a digital security shield that is proactive, anticipative, and able to neutralize threats even before they become apparent. This can be achieved through frequent technological updates and upgrades to the software, hardware, and network security measures, implementing comprehensive cybersecurity policies and regular training of employees and all stakeholders to adhere and comply with security practices.

CHAPTER 21   SECURING DIGITAL FOUNDATIONS: A POINT OF VIEW ON CYBERSECURITY IN HEALTHCARE

Zero Trust Model

The current healthcare cybersecurity landscape demands a "zero trust model" wherein the integrity of patient data is of utmost priority leaving no scope for any gaps or fallouts. Each and every user and device is thoroughly verified without a presumption of trust, thus preventing unauthorized access and also incidents of trojans and phishing incidents on highly secure patient data.

Future Threats and Mitigation Tactics

Ransomware and Advanced Malware

In the last decade, healthcare organizations have faced the most severe threat from ransomware attacks which not only impact them financially but also have the potential to compromise patient data as well as precious lives at risk owing to disruptions. As a result, healthcare organizations and their stakeholders are investing heavily in endpoint detection and response (EDR) systems and advanced threat protection (ATP) solutions to proactively detect and mitigate these threats.

Enhanced Encryption Practices

One of the simplest yet most effective solutions to prevent patient data compromise is end-to-end data encryption practice. This ensures that the data remains protected by implementing robust encryption standards both at transmission and end-user levels. Now let us look at the top four trends [15] that are impacting and likely to define how a robust digital security framework can protect critical healthcare data.

CHAPTER 21   SECURING DIGITAL FOUNDATIONS: A POINT OF VIEW ON CYBERSECURITY IN HEALTHCARE

Transformative Security Technologies

Blockchain for EHR

5G in Healthcare

Transformative Security Technologies

Predictive Analytics

VR & AR in Healthcare

Figure 21-7. *Four Key Transformative Security Technologies (Image Idea: Freepik.com)*

1. *Blockchain for Health Records*

   Blockchain technology in healthcare has evolved to provide an effective and efficient workflow to address the challenges of managing secure EHRs. Through its decentralized and immutable framework for safe and controlled data channels, blockchain ensures unmatched data integrity, privacy, and transparency. It addresses many of the current vulnerabilities intrinsic to traditional healthcare information systems. By enabling data to be stored in a secure and tamper-evident manner, blockchain technology not only protects information from unauthorized access but also facilitates the secure sharing of data across various healthcare stakeholders.

With the growing acceptance of blockchain technology in healthcare, future applications are expected to be more comprehensive and advanced. This technology has the potential to revolutionize how patient information is managed by streamlining the sharing of medical records in a secure environment, thus enhancing interoperability among different healthcare systems. For instance, a system secured using blockchain technology can enable a doctor or a hospital to access and view a patient's complete medical history and other PHIs with the patient's consent, and without any risk of data corruption or unauthorized alteration.

Blockchain could significantly speed up the administrative processes in healthcare settings by reducing the need for intermediaries and cutting down on the overhead costs associated with data management and exchange. The resulting outcomes would enhance the speed of data interchange as well as the cost efficiency of maintaining and security of sensitive patient health data. With widespread recognition of the benefits of blockchain and its adoption across the value chain, we can anticipate its integration into mainstream healthcare practices, setting new benchmarks for how patient data is handled in an increasingly digital world.

CHAPTER 21 SECURING DIGITAL FOUNDATIONS: A POINT OF VIEW ON CYBERSECURITY IN HEALTHCARE

2. *5G Connectivity for Real-Time Medical Applications*

5G technology has enabled faster connectivity speeds, low latency, and more capacity and connectivity to billions of devices. Its utilization in healthcare is more pronounced than we can envisage which could significantly bolster digital security measures within the industry. 5G offers more secure, real-time capabilities, needed to safeguard sensitive patient data with ultra-low latency and high bandwidth. 5G networks - and the higher speed and efficiency that can come with them - allow for near-instant synchronization and updating of security software, across devices and systems, to counteract cyber threats in real-time.

In addition, the use of 5G enables more sophisticated encryption protocols and secure communication channels, to ensure the security and integrity of the data exchanged between the devices and the healthcare providers. The capacity of a network bolstered by 5G connectivity, to process massive amounts of data in parallel, will also enable healthcare organizations to use advanced, real-time monitoring systems that can be programmed to perceive and respond to threats as they emerge.

Moreover, the entry of 5G expands the range of applications of AI and machine learning to be used in healthcare security. By analyzing the patterns in your network traffic, these technologies can catch what may be a new or potential cyberattack threat, thereby greatly increasing the protection with a level of security that is not only reactive but also adaptive. Leveraging 5G connectivity capabilities can increase the efficacy and impact of services provided by healthcare providers while ensuring a level of security far greater than the potential susceptibility of confidential patient information to the ever-increasing threat of cyberattacks.

3. *Virtual Reality (VR) and Augmented Reality (AR) in Healthcare*

Virtual reality (VR) and augmented reality (AR) are increasingly playing pivotal roles in bolstering the digital security features of healthcare systems. VR and AR make it easier for healthcare providers to receive customized and fine-tuned training and simulation, which is beneficial for improving the preparation of existing healthcare professionals in learning and combating cybersecurity incidents and their mitigation strategies. For example, VR can replicate cybersecurity breach scenarios in a healthcare setting, making it possible for IT and healthcare providers to execute response strategies in a simulated but real environment. This type of training is also indispensable for increasing the speed and efficiency of responses to real-world data breach incidents.

AR can also provide real-time overlay information to facilitate the examination and audit of healthcare systems, offering significant value to security operations. Envision a use case where AR can visually replicate and provide a demonstration of a security breach over a network architecture, thereby alerting operators and all stakeholders in learning and getting a first-hand experience of a potential security breach or the flow of proprietary data through systems and its encompassing vulnerabilities. This will enable operators to visually identify and proactively identify and address security incidents.

These technologies not only enhance the training and situational awareness of cybersecurity teams but also support the implementation of complex security protocols in an interactive manner that is easier to understand and apply. By integrating VR and AR into their digital security strategies, healthcare organizations can ensure a more robust defense mechanism, making their infrastructures less prone to cyberattacks and more resilient in safeguarding patient data.

4. *Predictive Analytics for Personalized Healthcare Management*

Predictive analytics is one of the crucial tools that the healthcare industry can implement as essential steps to fortify digital security measures. Using these patterns based on historical data, the technology does its best to predict and alert against probable security risks. Healthcare organizations that can successfully predict cyber threats before they are actualized can improve their security efforts proactively, customize their responses, and reduce the risks far more effectively.

For example, predictive analytics can detect uncharacteristic access to patient records or unusual network traffic - both key indicators of future data breaches and ransomware attacks. Healthcare systems can detect these early anomalies and take action to prevent malicious actions from accessing sensitive patient information or impairing critical infrastructure before that information is compromised or significant damage is done.

Furthermore, the integration of predictive analytics in healthcare cybersecurity measures will promote the development of better and sustained security protections. As these new data sources help machine learning systems learn from the dynamic and evolving threat environment, they will over time become even more accurate in predicting those behaviors and even more robust in defending against them.

In summary, predictive analytics thus becomes a strong tool, both directly preventing cyber threats and making healthcare digital security resilient, thus ensuring the systems are always one step ahead in the rapidly changing cyber threat landscape.

Case Examples and Empirical Evidence

In 2024, Adventist Health, a prominent healthcare organization, reported a significant data breach that exposed the sensitive information of over 70,000 individuals [16]. This incident served as a stark reminder of the critical importance of robust digital security measures in healthcare, particularly as organizations handle sensitive patient data on an increasingly digital platform.

CHAPTER 21   SECURING DIGITAL FOUNDATIONS: A POINT OF VIEW ON CYBERSECURITY IN HEALTHCARE

Figure 21-8. *Adventist Health Data Breach (Image Source:* https://classlawdc.com/2024/06/12/adventist-health-data-breach-investigation/)

The breach likely forced Adventist Health to conduct a thorough review and subsequent strengthening of its cybersecurity posture. This is a common response to such incidents, as organizations must learn from these experiences to prevent future breaches. This mirrors the actions taken by Universal Health Services following a similar breach, demonstrating a pattern in the healthcare industry's response to these threats.

In another instance, Panorama Eyecare, an eye care management company based in Fort Collins, Colorado, notified a staggering 377,911 individuals of a data breach that occurred in May 2023 [17]. The breach was claimed by LockBit ransomware, a notorious cyber threat that continues to plague the healthcare sector.

CHAPTER 21   SECURING DIGITAL FOUNDATIONS: A POINT OF VIEW ON CYBERSECURITY IN HEALTHCARE

Figure 21-9. *Panorama Eye Care Data Breach (Image Source: https://www.classaction.org/data-breach-lawsuits/panorama-eyecare-june-2024)*

These are just a couple of case examples emphasizing a critical requirement for comprehensive cybersecurity measures in the rapidly advancing digital healthcare landscape. Healthcare organizations must be mindful of these threats as they continue to grow and evolve. This awareness necessitates nurturing a cybersecurity knowledge base and talent and the adoption of new technological advances to protect patient information.

In addition to that, trust is very important for the digital healthcare ecosystem. Patients provide the most intimate details to healthcare providers; therefore, there is a duty to maintain this information securely for both regulatory and moral reasons. A strong digital security framework can support continued trust by showing you care for the privacy and security of the data.

CHAPTER 21 SECURING DIGITAL FOUNDATIONS: A POINT OF VIEW ON CYBERSECURITY IN HEALTHCARE

The Cybersecurity Ventures report's prediction about the increasing economic impact of cyber threats aligns with these cases. It serves as a sobering reminder of the potential financial implications of these threats, further emphasizing the need for proactive and robust digital security measures in healthcare.

The Cybersecurity Ventures report referenced in the New York Times in 2019 about the prediction for cybercrime costs is mirrored by the case studies illustrating the growing economic toll of such cyber threats. It serves as a stark example of the impact of the financial cost of such threats, thereby further reminding us of proactive and strong digital security rights for healthcare.

In conclusion, these real-life cases illustrate the multifaceted challenges of digital security in healthcare. They highlight the need for a dynamic approach, integrating advanced technologies, investing in cybersecurity talent, and continuously evolving strategies to safeguard patient data and maintain trust in the digital healthcare ecosystem. As we move further into the digital age, these challenges will only continue to grow, underscoring the importance of ongoing vigilance and adaptation in the face of an ever-evolving threat landscape.

Ultimately, these real-life case examples demonstrate the concerning and multidimensional threat landscape for digital security in healthcare. They illustrate the importance of giving a flexible response that integrates advanced technologies with reinvestments in cybersecurity talent and changing strategies to protect patient data and trust in the digital healthcare ecosystem. Cybersecurity experts foresee these challenges becoming more prevalent in the future, given the ever-evolving threat landscape, they simultaneously highlight that, as we make rapid strides into the digital age, parallelly these security challenges are going to evolve and present more challenges.

CHAPTER 21 SECURING DIGITAL FOUNDATIONS: A POINT OF VIEW ON CYBERSECURITY IN HEALTHCARE

Conclusion

The future of digital security in healthcare is not a static concept but rather a dynamic and multifaceted challenge that demands continuous evolution and adaptation. As we move further into the digital age, the integration of advanced technologies such as artificial intelligence (AI) and machine learning into traditional cybersecurity measures is becoming increasingly essential. These technologies offer the potential to predict and identify cyber threats before they can cause significant damage, thereby enhancing the resilience of the digital environment.

However, technology alone is not enough. The human element remains a critical factor in the cybersecurity equation. Therefore, comprehensive training programs for healthcare staff are imperative. These programs should aim to equip staff with the knowledge and skills to recognize and respond effectively to cyber threats. By fostering a culture of cybersecurity awareness, healthcare organizations can significantly reduce their vulnerability to cyber attacks.

As the digital landscape continues to evolve, so too must the strategies employed to protect it. This involves staying abreast of the latest cybersecurity trends and threats and continuously updating and refining security protocols and measures. The goal is to ensure the safety of critical healthcare data, which is fundamental to the provision of quality patient care. Maintaining the trust of patients and stakeholders is another crucial aspect of digital security in healthcare. Patients entrust healthcare providers with their most sensitive information, and there is a moral and legal obligation to protect this data. A robust digital security framework can help maintain this trust by demonstrating a commitment to data privacy and security.

Looking ahead, preemptive measures and strategic planning will play a pivotal role in safeguarding the digital frontier of healthcare. This involves not only defending against current cyber threats but also anticipating and preparing for future ones. By doing so, healthcare organizations can preserve the sanctity and integrity of patient care services, even in the face of an increasingly complex and challenging digital security landscape.

In conclusion, the future of digital security in healthcare is a multifaceted challenge that requires a dynamic approach, integrating advanced technologies, comprehensive staff training, continuous strategy evolution, and a strong commitment to data privacy and security. By meeting this challenge head-on, healthcare organizations can ensure a more secure and resilient digital future, thereby safeguarding the quality and integrity of patient care in the digital age.

References

[1] Extent of Magellan Health Ransomware Becomes Clear: More Than 364,000 Individuals Affected, By Steve Alder, July 1, 2020. https://www.hipaajournal.com/extent-of-magellan-health-ransomware-becomes-clear-more-than-364000-individuals-affected/

[2] Cyber Case Study: UVM Health Network Ransomware Attack by Kelli Young, Dec 6, 2021. https://coverlink.com/case-study/uvm-health-network-ransomware-attack/

[3] Cyber Risk in the Healthcare Industry by Syvanne Aloni, January 9, 2023. https://www.cyberinsuranceacademy.com/knowledge-hub/case-study/cyber-risk-in-the-healthcare-industry/

[4] Insider threat awareness and prevention with MS purview, Uncategorized / By Laura Young. https://levacloud.com/2024/02/29/insider-threat-awareness-with-ms-purview/

[5] The HIPAA Journal article by Steve Alder on January 4, 2024. https://www.hipaajournal.com/2023-healthcare-ransomware-attacks/

[6] Healthcare Dive article by Emily Olsen, on October 27, 2023. https://www.healthcaredive.com/news/healthcare-ransomware-costs-comparitech-77-billion/698044/

[7] Cyber Case Study: UVM Health Network Ransomware Attack by Kelli Young on Dec 6, 2021. https://coverlink.com/case-study/uvm-health-network-ransomware-attack/

[8] "7 Reasons for Yearly Technical Cybersecurity Audits in Healthcare" by Rob Abreu, Technical Solutions Architect at DAS Health. https://dashealth.com/blog/7-reasons-for-yearly-technical-cybersecurity-audits-in-healthcare/

[9] Cited—Antony, Anson & Thomas, Sanjo & Kunjachan Varghese, Titus & Padman, Vishnu. (2023). Ransomware Attacks on Healthcare Systems: Case Studies and Mitigation Strategies. 10.13140/RG.2.2.34192.17928. https://www.researchgate.net/publication/376514138_Ransomware_Attacks_on_Healthcare_Systems_Case_Studies_and_Mitigation_Strategies

[10] "2023 Ponemon Healthcare Cybersecurity Report." https://www.proofpoint.com/us/resources/threat-reports/ponemon-healthcare-cybersecurity-report

[11] "Train Employees And Cut Cyber Risks Up To 70 Percent" by Stu Sjouwerman. https://blog.knowbe4.com/train-employees-and-cut-cyber-risks-up-to-70-percent#:~:text=When%20they%20are%20exposed%20to,as%20much%20as%2070%20percent.

[12] The February 2024 cyberattack on Change Healthcare: A case study in healthcare cybersecurity vulnerabilities. WO3497-Whitepaper-41_The-February-2024-Cyberattack-on-Change-Healthcare_-A-Case-Study-in-Healthcare-Cybersecurity-Vulnerabilities-2024_04_10.pdf (caplinehealthcaremanagement.com)

[13] Ransomware Attacks Surge in 2023; Attacks on Healthcare Sector Nearly Double, February 28, 2024. *https://www.dni.gov/files/CTIIC/documents/products/Ransomware_Attacks_Surge_in_2023.pdf*

[14] Ransomware_Attacks_Surge_in_2023.pdf (dni.gov)—2023 Health Cybersecurity Annual Threat Report published on March 6, 2023, Threat Intelligence, White Papers. https://h-isac.org/2023-health-cybersecurity-annual-threat-report/

CHAPTER 21   SECURING DIGITAL FOUNDATIONS: A POINT OF VIEW ON CYBERSECURITY IN HEALTHCARE

[15] "Digital Transformation in Healthcare," Nov 16, 2023. https://www.aissel.com/blog/Digital-Transformation-in-Healthcare

[16] Adventist Health West reports major data breach—The Sun-Gazette Newspaper (thesungazette.com). June 11, 2024.

[17] "Eyecare company suffers 377K—record data break," Health IT Security, June 11, 2024. https://healthitsecurity.com/news/eye-care-company-suffers-377k-record-data-breach

PART VI

Future Trends in Cybersecurity

Chapter 22: Future Trends in Digital Security

- **Emerging Technologies and Methodologies:** Explores emerging technologies and methodologies that are likely to shape the future of digital security, including artificial intelligence, machine learning, and blockchain.

- **Quantum Cryptography and AI in Security:** Delves into the potential impacts of quantum computing on encryption and the growing role of AI in cybersecurity, both as a tool for defenders and a weapon for attackers.

CHAPTER 22

Future Trends in Digital Security

Author:
Anirudh Khanna

Future Trends and Anticipations in Digital Security

The current philosophical debate in matters related to digital security revolves around the need to uphold state security laws like the EU's 2016 General Data Protection Regulation and preserve the right to privacy and trust attached to the principle of end-to-end encryption. Notably, considering all the innovative developments expected, the major challenge related to these recurring themes in the future of cybersecurity and digital security is how to deal with the global [1]. Emerging technologies like blockchain and artificial intelligence should be considered when elaborating on an agenda for the future of digital security.

Artificial Intelligence and Machine Learning

AI is integral in today's digital society, capturing a broad set of activities that enable technology to improve healthcare, transportation, finance, and security. According to Camacho [2], the sub-field of machine learning has produced patterns that can make impactful decisions, informed by a higher number of data sources than any human, such that these critical sectors of society are taking a newfound trust bias to these methodological outputs. Machine learning-based cybersecurity products like artificial intelligence detection software, exemplified by products like *Elastio*, are a method of finding security threats quickly and precisely [3]. Automated machines and systems that learn to identify threats from various data make it easier for cybersecurity staff to focus their exertions. Several major cybersecurity corporations now use digital security strategies and devices based on AI.

Zero Trust Architecture

Zero trust architecture (ZTA)—also known as zero trust networking or zero trust network access (ZTNA)—is a cybersecurity model developed by John Kindervag in 2010 [5]. ZTA is promoted by a series of individuals and organizations, notably Forrester Research. The architecture is based on creating zones by utilizing the network surface as an attack vector. It requires the assignment, verification, and explicit validation of strict identity and access requirements for every person, device, and network attempting to access assets on the network. Through zero trust principles, an organization has no physical network perimeter. It considers every connection, user, device, and system that attempts to enter the network as a threat that could adversely impact the network, regardless of whether it is internal, as demonstrated by Figure 22-1. Such developments will continue to be integral in the future of digital security because every person, access point, network, and networked device are vulnerabilities that must be continuously interrogated for cyber threats.

CHAPTER 22   FUTURE TRENDS IN DIGITAL SECURITY

Figure 22-1. *The Zero Trust Architecture of Cybersecurity [6]*

Edge Computing Strategies

Among its many features, edge computing promises to extract data only generated by events or people, sometimes translated as time-series data [7]. Therefore, edge computing digital security relies on allowing employees and operators access to resources within systems and networks relevant to the tasks and queries they or their verified devices make. One critical challenge posed by cybersecurity threats is the constant networking, which predisposes all data and systems to threats. However, with edge computing, specific resources are walled away from potential and actual threats [8]. Also, an emerging threat in digital security is sorting data types within online systems and training security and access control algorithms to deliver insights into what is going on during a potential attack, preparing the entire system for digital resilience through machine learning processes.

CHAPTER 22 FUTURE TRENDS IN DIGITAL SECURITY

Cloud Computing Strategies

Cloud computing will be relevant to digital security in the future because it will provide access to several resources, including applications, infrastructure, or services, using an application programming interface. The data is stored and processed on remote servers rather than locally used, private computers [8]. The scalability of cloud computing is compelling because it offers opportunities to move faster and scale up without additional capital expenditure. Therefore, cloud computing benefits undercapitalized businesses that aspire toward digital resilience and digital security by accessing formidable cybersecurity functionalities such as those provided by Amazon Web Services (AWS) to its client infrastructures [10]. These factors demonstrate why many people and organizations prefer cloud computing, as shown in Figure 22-2, where several functionalities, such as security, network optimization, mobile applications, and storage, are combined. Such operations become their preferred form of IT procurement and deployment. Cloud computing is a significant step forward in digital security for humans, software, and hardware because utilizing and sharing online resources can bring various solutions and protections to bear on any current and future cyber threats.

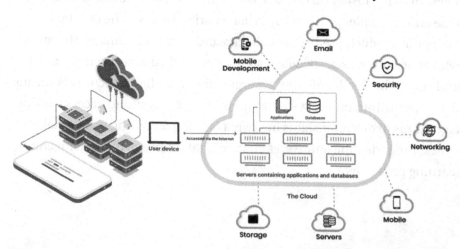

Figure 22-2. *Amazon Web Services Cloud Computing Operations [9]*

CHAPTER 22   FUTURE TRENDS IN DIGITAL SECURITY

Quantum Computing in Data Encryption

A critical difference between quantum and classical mechanics is that a quantum system is only considered to have specified entirely properties once a measurement of those properties is made. Consequently, a quantum system can exist in a superposition of states, but once a measurement is made, the superposition is projected onto one of those states [11]. Only the result of the measurement can be known. One can use quantum mechanical phenomena to perform mathematical calculations currently infeasible using classical means, as evidenced by quantum computing [12]. A quantum computer performs computation by manipulating these superpositions of states of the various devices in a quantum web until it achieves the final result. Such cutting-edge computing innovations promise impressive data security capabilities that could benefit information technology. However, quantum mechanics challenges the security of classical cryptology [13]. Shor's algorithm, an algorithm in quantum computing, would break public key cryptography. However, ongoing research and innovations point to formidable encryption and digital security innovations shortly [41].

The Role of Encryption in Futuristic Digital Security
Post-quantum Encryption

NIST has been running a process to solicit, select, and standardize new post-quantum digital signature and public key encryption schemes for the internet [14]. Unfortunately, not all quantum-resistant encryption schemes provide all the benefits of current encryption. Also, the sender and recipient can quickly produce and consume public keys, anyone can encrypt messages, and the decryptor can determine the sender and whether error correction is possible.

CHAPTER 22   FUTURE TRENDS IN DIGITAL SECURITY

In the future of digital security, encryption will continue to play a critical role. As the advent of advanced quantum computers looms, it will be more critical than ever to update widely used encryption protocols like Advanced Encryption Standard (AES) and Triple DES (Data Encryption Standard). Shortly, a quantum computer—should it ever come online—could break current public essential encryption methods [13]. Researchers are contemplating a move to longer encryption keys at 128 keys compared to the standard 64 keys. That will be resistant to quantum computer attacks. If such computers become viable, they can efficiently solve the factoring problems, and discrete logarithms underlie widely deployed public essential encryption methods. This obscure mathematical assertion can be reduced to an easily understandable claim: Quantum computers would be able to break our digital encryption.

Homomorphic Encryption

Also, the role of post-quantum encryption technologies must be addressed, considering how close the state of technology is to fully functional quantum computers. Unfortunately, quantum computing has the theoretical capability to render many encryption algorithms impotent, meaning the future of digital security is tied to developments anticipated in post-quantum cryptography [15]. Fully Homomorphic Encryption (FHE) is an encryption technique that allows a party to perform arbitrary computations on ciphertext and is rapidly gaining attention from the fields of cryptographic and machine learning research [16]. The goal is to allow computation to be delegated to an untrusted party without sharing the underlying data, enabling the construction of cloud-based services that accept encrypted inputs and return the encrypted results. When discussing these novel encryption schemes, we will focus on two major themes derived from the paper: Faster homomorphic linear transformations.

In the early 2010s, Shai Halevi and Victor Shoup released a new encryption scheme called the BGV encryption scheme, which implemented a form of encryption operation called homomorphic encryption [17]. FHE is still in development, but the BGV scheme was significant due to its idea that you could use the basic operations of Fully Homomorphic Encryption (FHE) to perform other "higher order" operations, such as simple addition and multiplication on encrypted plaintext data [15]. This was a big step forward in the cryptographic research community, and it has taken a decade for this research to reach its full potential. These developments have set the stage for ongoing innovations and research into FHE as a future developmental pillar in digital security.

Encrypted Machine Learning

Secure multiparty computation (MPC) studies distributed, multiparty computation, which has building blocks of encrypted machine learning (EML). In the context of future digital encryption, privacy-preserving techniques disallow the public auditor from learning any private information while computing the function of interest [18]. Encrypted machine learning strategies will complement the future of digital security by using techniques and protocols to perform computational tasks while preserving the privacy and security of data. This emerging field aims to find ways to enable collaborative data analysis and machine learning models without exposing sensitive information to external threats [19]. By combining encryption methods with machine learning algorithms, secure computation and encryption with encrypted machine learning offer a powerful solution for protecting data confidentiality and privacy. These techniques allow multiple parties to jointly compute encrypted data without revealing individual inputs to each other or any third party [20]. With ongoing advancements in the field, secure computation and encryption with encrypted machine learning hold great promise for many applications, including healthcare, finance, and data analytics.

Biometric Authentication and Identity Management
AI-Enabled Multimodal Biometrics

A promising method to enhance biometric systems is to combine two or more biometrics like fingerprints with facial scans with artificial intelligence for recognition performance and error detection. Multimodal identity management is also referred to as a hybrid biometric system, and by dramatically reducing the error rates, the overall system's security increases disproportionately as the errors of the combined systems are inversely proportional to the error in the constituent systems [22]. Moreover, multimodal biometrics alleviates the problem of noisy data. By combining a limited error rate biometric characteristic with a complementary biometric characteristic that suffers from independent error types, an accuracy rate can be attained significantly better than either characteristic alone [23]. However, while it may be difficult for an attacker to statistically determine a unique personal ID code from a single data source and a different unique personal ID code from another source, the simple mathematical accumulation of low individual performance rates makes the system more susceptible to attack.

Nevertheless, due to the substantial security value of multimodal biometrics, several countries, including Poland and India, have established policy-supporting infrastructures to prepare the groundwork for multimodal biometrics adoption [24]. Current multimodal biometrics use voice, iris, and face. In the future, advanced biometrics will include gait, odor, soft-biometrics (e.g., gender, height, weight), blood group, DNA, hand veneer pattern, ear shape, and ocular vasculature, as demonstrated by Figure 22-3, where several functionalities come together as a unified operation.

CHAPTER 22 FUTURE TRENDS IN DIGITAL SECURITY

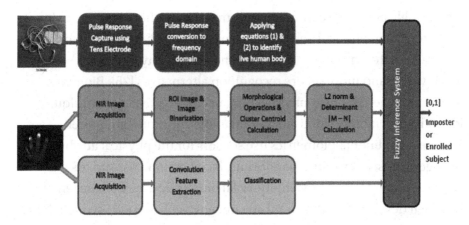

Figure 22-3. *A Multimodal Biometrics System Combined with Essential AI Functions [21]*

Human–Machine Adaptation

Using continuous biometric recognition adaptations, the future evolution of automatic recognition systems must handle users who move through security environments, typically in the direction of registration (enrollment) and identification, utilizing further data whenever signals do not fall into any of the recognition streams prepared but for whose existence, at a time t, mathematical fusion did not connect [24]. This adaptive property is beneficial when trying to contain visible and invisible threats at intrusion points when the setup of detection and liveness systems that use multimodal biometrics is incrementally activated due to changes in security risk [25]. Also, when hazardous events are predicted at a given instant, multimodal biometrics sensors can gather more anthropometric details concerning subject surveillance status.

Furthermore, various personal security parameters and authorizations are often linked to a user's capability to draw upon predefined, non-disclosed recognition strategies. Hence, the strategy used at a given instant in decision-making processes is also related to accessing or using increasingly pervasive controlled resources [39]. Other systems can be exploited using physical and cognitive multimodal biometrics, which can pivot (when the amplitude of instrumental sensitivity plummets) onto less invasive and more secure recognition alternatives [26]. Biometric access management uses physical human characteristics as a unique identification method for users to gain authentication for required facilities. Traditional biometrics use sensors for the physical attributes of fingerprints, palmprints, handprints, facial landmarks, and iris and vascular patterns. Recently, multimodal biometric access management has been developed with several advantages, such as increased reliability, improved accuracy, and the provision of a more extensive, comprehensive, and rugged database than the unimodal system.

Blockchain Technology and Security Applications

The pervasiveness of blockchain from concept to development is unprecedented and could also contribute to the ongoing developments of digital security. The range of uses of blockchain technology is growing in commercial applications, presenting cybersecurity challenges unique to this application. This fundamental technology is permeating virtually all market variables and may have the potential to fundamentally change global interaction at both a macro and microeconomic level [27]. Blockchain is a distributed ledger that is secure, trustworthy, and transparent due to its proposed technologies. It is a chronological chain encrypted with cryptographic solid techniques, in which block elements are distributed. Its properties make it tamperproof, i.e., one cannot change previously entered records or records after some time [4]. Authorized participants may be able to access and exchange shared data distributed across multiple channels, which can be adequately written and controlled by a combination of a distributed ledger with interfaces

CHAPTER 22   FUTURE TRENDS IN DIGITAL SECURITY

encoded into smart contracts. New data are constantly added to the ledger in a perpetual, process-driven, timestamped series of time-ordered transactions.

How blockchain technology locks user and transaction information or data using variable encryptions that only authorized users can access and use is usable in the future of digital security. Therein, data security firms could utilize such methodologies to ensure that decentralized encryption software that must be authenticated by all relevant, authorized, and authenticated parties could be helpful in futuristic data protection and device access software [28]. Furthermore, decentralized blockchain technology allows no single entity to have higher scrutiny than most parties involved in each transaction. Mutability and irrefutability are two key features that add an extra layer of protection to the blockchain, as every transaction has to be agreed upon by each party included and approved [29]. The data and activities are traceable from beginning to end, as are the transaction credentials; therefore, blockchain right back provides a complete audit trail.

Cybersecurity in the Internet of Things (IoT)

While new mechanisms may prompt IoTs to assume communication networks are secure by design, novel PIT applications are deployed in non-typical network infrastructures, such as wireless ad hoc or P2P topologies, that can benefit from these new secure communications features. Dynamically formed RF channels enable secure communications under silence (i.e., listen-before-talk) regimes, allowing highly secure and energy-efficient in-device networking that looks set to revolutionize the design of the Internet of Things [30]. As new IoT applications emerge, new ways to encapsulate security and trust mechanisms within the devices become a significant challenge. In particular, the miniaturization of devices caused by integration with different production technologies demands dedicated physical and cryptographic security primitives that can be widely diffused to form practical in-device security solutions [31]. The current IRDS calls for new advances in physical-layer key generation and protection as critical enablers for secure communications [40]. New initiatives are thus expected to provide novel encryption and steganographic techniques that guarantee safe and privacy-aware communications without adding hardware complexity [32]. Optical steganography is gaining momentum due to the wiretap security of optical fiber transmissions. Similarly, the next generation of tagless RFID technology is also emerging as an advanced security mechanism.

Security in a Remote Work Environment

With the outbreak of the COVID-19 pandemic in 2019, practically every organization outside of healthcare had to close its physical doors. People who had never previously worked remotely found themselves using personal devices that needed to be more secure and supported. The widespread adoption of work-from-home practices will likely happen over the next few years. What the COVID-19 crisis did was accelerate that process exponentially. Companies realized the productive potential of remote work and the need to invest less in physical infrastructure [33]. Now that the technology and processes are still in transition and testing is suddenly official and in use, the scope of applications that allow employees to work the best for them may completely thwart the security precautions to protect company assets.

At the beginning of 2020, many organizations considered themselves significantly prepared for remote work. Employees had company-approved devices, complete with firewalls and antivirus programs. They accessed their systems through VPNs and even logged in through multi-factor authentication. Remote work was a known reality, as was business continuity. Unfortunately, the added autonomy among certain employees working in high-risk departments like IT, engineering, HRM, and legal means their devices are particularly susceptible to hacking, ransomware, and phishing attacks in less secure work environments [34]. This digital security threat gets compounded when they bring their compromised network devices, like laptop computers, back to the office, enabling ransomware and viruses to infect the entire network. Therefore, remote work practices pose a unique challenge to the future of digital security.

Shortage of Skilled Talent in Digital Security

The skills deficit in digital security has an impact, as evidenced by the number of attacks that enterprises and governments are increasingly suffering, forcing the need to invest in solutions and protection mechanisms to address the current shortage of 3.4 million cybersecurity professionals [42]. There is a massive shortage of digital security personnel in various industries, as demonstrated in Figure 22-4. This increased market demand is critical—it increases the industry's economic footprint, creates new opportunities, and stimulates the competition that ensures the continuous development of increasingly effective digital security solutions, technologies, and techniques [35]. Without digital security, many of the anticipated benefits of digital transformation, including creative, financial, and operational efficiency, will be compromised. Addressing the shortfalls in current digital security skills is crucial to prevent future attacks and establish a level playing field.

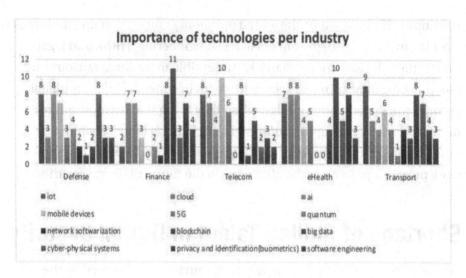

Figure 22-4. *Shortage of Digital Security and Cybersecurity Talent Per Industry in the EU as of 2021 [38]*

To address the skills gap for digital security in the future, one must acknowledge a mismatch between expected and current realities. The fast-paced evolution of technologies and the gamut of changes in the types and nature of threats call for synchronizing the pace of skills development and deployment. The nature of exposure also needs to be questioned. The recent surge in training and assessment programs in digital forensics and cybersecurity in significant colleges and universities worldwide demonstrates the need for these subjects, not the actual outcomes or delivery to the job market for improving digital resilience to meet the upsurge of malware and cyberattacks.

The demand for security professionals will continue to rise steeply across job sectors. Information from the Bureau of Security and Investigative Services within the California Department of Consumer Affairs indicates that close to four million unfilled cybersecurity jobs exist in the United States alone. The crisis is not confined to the United States. A McKinsey-commissioned study reports a similar trend on a global scale.

CHAPTER 22　FUTURE TRENDS IN DIGITAL SECURITY

The security skills crisis has become a genuine concern for organizations worldwide. A 2019 report by ISC highlights that the cybersecurity talent gap is immense [36]. By 2022, the report estimates the shortfall to be over four million professionals worldwide. Fixing the talent gap calls for careful consideration on multiple fronts. A combination of systematic interventions, as well as market and policy corrections, are required. This section explores the factors contributing to the scarcity and mismatch of needed skills [37].

Notably, many universities and technical colleges are engaged in private-public partnerships to develop new and future curricula for digital security specialists for current market needs and the anticipated upsurge of demand expected in the future. Let us now analyze future digital and cybersecurity measures in healthcare as a case example.

The Digital Transformation Landscape in Healthcare
Consumer-Centric Approach

Digital capabilities are becoming essential tools for healthcare systems eager to transform consumer relationships. Healthcare is becoming more consumer-focused, and organizations are rethinking what digital experiences can mean for the patient experience. This shift requires designing processes from the consumer's perspective, building trust and loyalty, and introducing innovative forms of digital care delivery. However, this increased digital interaction also expands the attack vectors, making protecting confidential patient data with secure measures more imperative.

Interim Milestones and Value Measurement

A digital transformation journey involves setting interim milestones and proving the value of digital initiatives. Health systems frequently assess these milestones to measure ROI and evolve their security strategies in response to the ever-evolving nature of threat risks. The health system must adopt these strategies of continuous improvement.

Talent and Data Challenges

The success of digital transformation heavily relies on the quality and availability of talent. Healthcare organizations prioritize recruiting and developing skilled professionals who analyze digital technologies and cybersecurity complexities. Eliminating data interoperability challenges and defining the right key performance indicators (KPIs) will ensure data security and governance.

As healthcare systems move to be consumer-centric, a robust digital security setup becomes more critical [41]. Implementing digital solutions to enrich patient experiences is accompanied by increased exposure to cyber threats. This has, in turn, led to the implementation of high-tech security measures for the safe data storage of their patients and the stability of their online systems. As they navigate various stages of their digital transformation journey, continuously evaluating and fortifying their cybersecurity frameworks is essential to safeguard against evolving threats and maintain consumer trust.

Summary

The chapter details the future digital security trends in the context of cybersecurity developments. The five domains involved in the treatise are artificial intelligence/machine learning, zero trust architecture, edge and cloud computing, and quantum computing. In addition, the role of encryption innovations in the future of digital security comes under inspection alongside biometrics and identity management. Also, the Internet of Things has a role to play in this development, and blockchain technology's contribution to digital security becomes evident. Eventually, the risks of remote work on digital security and the gaps that exist in talent intended for digital security also benefit from the discussion.

References

[1] S. Fischer-Hübner *et al.*, "Stakeholder perspectives and requirements on cybersecurity in Europe," *Journal of Information Security and Applications*, vol. 61, p. 102916, Sep. 2021, doi: https://doi.org/10.1016/j.jisa.2021.102916.

[2] Nicolas Guzman Camacho, "The Role of AI in Cybersecurity: Addressing Threats in the Digital Age," *Journal of Artificial Intelligence General science (JAIGS) ISSN 3006-4023*, vol. 3, no. 1, pp. 143–154, Mar. 2024, doi: https://doi.org/10.60087/jaigs.v3i1.75.

[3] M. Waqas, S. Tu, Z. Halim, S. U. Rehman, G. Abbas, and Z. H. Abbas, "The role of artificial intelligence and machine learning in wireless networks security: principle, practice and challenges," *Artificial Intelligence Review*, vol. 55, no. 7, Feb. 2022, doi: https://doi.org/10.1007/s10462-022-10143-2.

[4] S. Singh, I.-H. Ra, W. Meng, M. Kaur, and G. H. Cho, "SH-BlockCC: A secure and efficient Internet of things smart home architecture based on cloud computing and blockchain technology," *International Journal of Distributed Sensor Networks*, vol. 15, no. 4, p. 155014771984415, Apr. 2019, doi: https://doi.org/10.1177/1550147719844159.

[5] Y. He, D. Huang, L. Chen, Y. Ni, and X. Ma, "A Survey on Zero Trust Architecture: Challenges and Future Trends," *Wireless Communications and Mobile Computing*, vol. 2022, pp. 1–13, Jun. 2022, doi: https://doi.org/10.1155/2022/6476274.

[6] J. Martinez, "What is Zero Trust? Implementation, Best Practices and More | strongDM," *discover.strongdm.com*, Jun. 10, 2024. https://www.strongdm.com/zero-trust

[7] G. Carvalho, B. Cabral, V. Pereira, and J. Bernardino, "Edge computing: current trends, research challenges and future directions," *Computing*, vol. 103, no. 5, Jan. 2021, doi: https://doi.org/10.1007/s00607-020-00896-5.

[8] X. Li, T. Chen, Q. Cheng, S. Ma, and J. Ma, "Smart Applications in Edge Computing: Overview on Authentication and Data Security," *IEEE Internet of Things Journal*, vol. 8, no. 6, pp. 4063–4080, Mar. 2021, doi: https://doi.org/10.1109/jiot.2020.3019297.

[9] Signity Solutions, "AWS Cloud Architecture Services - Signity Solutions," www.signitysolutions.com, 2024. https://www.signitysolutions.com/aws-cloud-architect-services (accessed Jun. 15, 2024).

[10] S. El Kafhali, I. El Mir, and M. Hanini, "Security Threats, Defense Mechanisms, Challenges, and Future Directions in Cloud Computing," *Archives of Computational Methods in Engineering*, vol. 29, no. 1, Apr. 2021, doi: https://doi.org/10.1007/s11831-021-09573-y.

[11] F. Raheman, "The Future of Cybersecurity in the Age of Quantum Computers," *Future Internet*, vol. 14, no. 11, p. 335, Nov. 2022, doi: https://doi.org/10.3390/fi14110335.

[12] D. Denning, "Is Quantum Computing a Cybersecurity Threat?," *American Scientist*, vol. 107, no. 2, p. 83, 2019, doi: https://doi.org/10.1511/2019.107.2.83.

[13] V. Hassija, V. Chamola, A. Goyal, S. S. Kanhere, and N. Guizani, "Forthcoming applications of quantum computing: peeking into the future," *IET Quantum Communication*, vol. 1, no. 2, Nov. 2020, doi: https://doi.org/10.1049/iet-qtc.2020.0026.

[14] H. Alyami *et al.*, "Analyzing the Data of Software Security Life-Span: Quantum Computing Era," *Intelligent Automation & Soft Computing*, vol. 31, no. 2, pp. 707–716, 2022, doi: https://doi.org/10.32604/iasc.2022.020780.

[15] H. Habri, A. Chillali, and A. Boua, "KEY MATRICES IN FULLY HOMOMORPHIC ENCRYPTION," *JP Journal of Algebra, Number Theory and Applications*, vol. 54, pp. 35-50, Feb. 2022, doi: https://doi.org/10.17654/0972555522014.

[16] A. Alharbi, H. Zamzami, and E. Samkri, "Survey on Homomorphic Encryption and Address of New Trend," *International Journal of Advanced Computer Science and Applications*, vol. 11, no. 7, 2020, doi: https://doi.org/10.14569/ijacsa.2020.0110774.

[17] B. Alaya, L. Laouamer, and N. Msilini, "Homomorphic encryption systems statement: Trends and challenges," *Computer Science Review*, vol. 36, no. 6, p. 100235, May 2020, doi: https://doi.org/10.1016/j.cosrev.2020.100235.

[18] M. Wazid, A. K. Das, V. Chamola, and Y. Park, "Uniting cyber security and machine learning: Advantages, challenges and future research," *ICT Express*, vol. 8, no. 3, Apr. 2022, doi: https://doi.org/10.1016/j.icte.2022.04.007.

[19] Z. Wang, K. W. Fok, and V. L. L. Thing, "Machine learning for encrypted malicious traffic detection: Approaches, datasets and comparative study," *Computers & Security*, vol. 113, p. 102542, Feb. 2022, doi: https://doi.org/10.1016/j.cose.2021.102542.

[20] H. Fang and Q. Qian, "Privacy Preserving Machine Learning with Homomorphic Encryption and Federated Learning," *Future Internet*, vol. 13, no. 4, p. 94, Apr. 2021, doi: https://doi.org/10.3390/fi13040094.

[21] S. A. Haider, Y. Rehman, and S. M. U. Ali, "Enhanced Multimodal Biometric Recognition Based upon Intrinsic Hand Biometrics," *Electronics*, vol. 9, no. 11, p. 1916, Nov. 2020, doi: https://doi.org/10.3390/electronics9111916.

[22] R. Ryu, S. Yeom, S.-H. Kim, and D. Herbert, "Continuous Multimodal Biometric Authentication Schemes: A Systematic Review," *IEEE Access*, vol. 9, pp. 34541–34557, 2021, doi: https://doi.org/10.1109/access.2021.3061589.

[23] M. Hammad, Y. Liu, and K. Wang, "Multimodal Biometric Authentication Systems Using Convolution Neural Network Based on Different Level Fusion of ECG and Fingerprint," *IEEE Access*, vol. 7, pp. 26527–26542, 2019, doi: https://doi.org/10.1109/access.2018.2886573.

[24] V. Rajasekar *et al.*, "Efficient Multimodal Biometric Recognition for Secure Authentication Based on Deep Learning Approach," *International Journal on Artificial Intelligence Tools*, no. 3, Feb. 2023, doi: https://doi.org/10.1142/s0218213023400171.

[25] J.-P. A. Yaacoub, H. N. Noura, O. Salman, and A. Chehab, "Robotics Cyber security: vulnerabilities, attacks, countermeasures, and Recommendations," *International Journal of Information Security*, vol. 21, no. 21, Mar. 2021, doi: https://doi.org/10.1007/s10207-021-00545-8.

[26] C. U. Akpuokwe, A. O. Adeniyi, and S. S. Bakare, "LEGAL CHALLENGES OF ARTIFICIAL INTELLIGENCE AND ROBOTICS: A COMPREHENSIVE REVIEW," *Computer Science & IT Research Journal*, vol. 5, no. 3, pp. 544–561, Mar. 2024, doi: https://doi.org/10.51594/csitrj.v5i3.860.

[27] Z. Xiong, Y. Zhang, D. Niyato, P. Wang, and Z. Han, "When Mobile Blockchain Meets Edge Computing," *IEEE Communications Magazine*, vol. 56, no. 8, pp. 33–39, Aug. 2018, doi: https://doi.org/10.1109/mcom.2018.1701095.

[28] S. Demirkan, I. Demirkan, and A. McKee, "Blockchain Technology in the Future of Business Cyber Security and Accounting," *Journal of Management Analytics*, vol. 7, no. 2, pp. 1–20, Feb. 2020, doi: https://doi.org/10.1080/23270012.2020.1731721.

[29] S. Singh, A. S. M. Sanwar Hosen, and B. Yoon, "Blockchain Security Attacks, Challenges, and Solutions for the Future Distributed IoT Network," *IEEE Access*, vol. 9, pp. 1–1, 2021, doi: https://doi.org/10.1109/access.2021.3051602.

[30] T. Gebremichael *et al.*, "Security and Privacy in the Industrial Internet of Things: Current Standards and Future Challenges," *IEEE Access*, vol. 8, pp. 152351–152366, 2020, doi: https://doi.org/10.1109/access.2020.3016937.

[31] N. Jhanjhi, M. Humayun, and S. N. Almuayqil, "Cyber Security and Privacy Issues in Industrial Internet of Things," *Computer Systems Science and Engineering*, vol. 37, no. 3, pp. 361–380, 2021, doi: https://doi.org/10.32604/csse.2021.015206.

[32] A. E. Omolara *et al.*, "The internet of things security: A survey encompassing unexplored areas and new insights," *Computers & Security*, vol. 112, p. 102494, Jan. 2022, doi: https://doi.org/10.1016/j.cose.2021.102494.

[33] F. Malecki, "Overcoming the security risks of remote working," *Computer Fraud & Security*, vol. 2020, no. 7, pp. 10–12, Jul. 2020, doi: https://doi.org/10.1016/S1361-3723(20)30074-9.

[34] A. Aloisi and V. De Stefano, "Essential jobs, remote work and digital surveillance: addressing the COVID-19 pandemic panopticon," *International Labour Review*, vol. 161, no. 2, pp. 289–314, Jun. 2021, doi: https://doi.org/10.1111/ilr.12219.

[35] S. Furnell, "The Cybersecurity Workforce and Skills," *Computers & Security*, vol. 100, p. 102080, Oct. 2020, doi: https://doi.org/10.1016/j.cose.2020.102080.

[36] T. Karaboğa, Y. Gürol, C. M. Binici, and P. Sarp, "Sustainable Digital Talent Ecosystem in the New Era: Impacts on Businesses, Governments and Universities," *Istanbul Business Research*, Jan. 2021, doi: https://doi.org/10.26650/ibr.2020.49.0009.

[37] J. Frenk *et al.*, "Health Professionals for a New century: Transforming Education to Strengthen Health Systems in an Interdependent World," *The Lancet*, vol. 376, no. 9756, pp. 1923–1958, Dec. 2020.

[38] B. J. Blažič, "The cybersecurity labour shortage in Europe: Moving to a new concept for education and training," *Technology in Society*, vol. 67, p. 101769, Nov. 2021, doi: https://doi.org/10.1016/j.techsoc.2021.101769.

[39] Ernest, "What Is The Future of Digital Security," *Security Boulevard*, Sep. 12, 2023. https://securityboulevard.com/2023/09/what-is-the-future-of-digital-security/#:~:text=The%20future%20of%20digital%20security%20will%20require%20a%20multifaceted%20approach (accessed Jun. 15, 2024).

[40] Cyber Security, "The Future of Cybersecurity Emerging Technologies," *Intone Networks*, Jun. 12, 2024. https://intone.com/the-future-of-cybersecurity-emerging-technologies/#:~:text=Cybersecurity%20emerging%20technologies%2C%20whether%20in (accessed Jun. 15, 2024).

[41] D. Joseph *et al.*, "Transitioning organizations to post-quantum cryptography," *Nature*, vol. 605, no. 7909, pp. 237–243, May 2022, doi: https://doi.org/10.1038/s41586-022-04623-2.

[42] K. Xie, "Cybersecurity: Close the skills gap to improve resilience," *World Economic Forum*, Feb. 01, 2023. https://www.weforum.org/agenda/2023/02/cybersecurity-how-to-improve-resilience-and-support-a-workforce-in-transition/

Index

A

ABAC, *see* Attribute-Based Access Control (ABAC)
Access control/authentication approaches, 70
   Attribute-Based Access Control (ABAC), 61
   cybersecurity, 54
   discretionary access control (DAC), 58
   implementation process, 62
      implementation, 64
      least privilege principle, 63
      monitoring/logging, 63
      regular audits, 62
      requirements, 62
      review access, 63
      user training, 63
   levels of, 58
   mandatory access control (MAC), 59, 60
   role-based access control (RBAC), 60, 61
   scalability, 59
Active Directory (AD), 110
AD, *see* Active Directory (AD)
Adjunct detections
   AV/EDR system, 145
   implementation/benefits, 148
   proactive security posture, 147
   signatures/behaviors, 147
   vigilance/responsive, 147
Advanced encryption standard (AES), 48, 371, 486
Advanced persistent threat (APT), 27, 85, 106, 180
Advanced threat protection (ATP), 464
AES, *see* Advanced encryption standard (AES)
AI, *see* Artificial intelligence (AI)
APT, *see* Advanced persistent threat (APT)
ATP, *see* Advanced threat protection (ATP)
Amazon Web Services (AWS), 369, 484
Anti-money laundering, 269
Antivirus (AV), 146–148
AR, *see* Augmented reality (AR)

INDEX

Artificial intelligence (AI), 9, 16, 28–30, 374, 481
    adversarial machine learning, 28
    automation/improvement, 28
    data poisoning, 29
    detection techniques, 151
    fake videos/audio, 29
    malware, 29
    model inversion/extraction, 29
    social engineering/phishing activities, 29
    vulnerabilities, 30
Artificial Intelligence (AI) backup/restore, 223
Asymmetric encryption model, 51, 52
Attackers
    ransomware, 259 (*see also* Ransomware)
    cyber criminals, 255
    mobile malware, 255
    United States/Europe, 256–259
Attribute-Based Access Control (ABAC), 61
Augmented reality (AR), 468
Authentication
    access control (*see* Access controls/authentication)
    biometric authentication, 57, 58, 82, 83
    biometric methods, 77
    categories, 54
    component, 79
    components, 54
    cybersecurity, 77, 78
    definition, 76, 77
    digital signatures/digital certificates, 77
    effective method, 78
    identification, 79
    implementation multifactor, 91, 92
    inherence factors, 91
    knowledge factors, 91
    multifactor, 76
    multifactor authentication (MFA), 83, 84
    password-based authentications, 80, 81
    password policies, 55–57
    possession factors, 91
    regular password updates, 89, 90
    resources, 75
    strategies, 87
    strong password policies, 88, 89
    two-factor authentication, 81
    two-step verification, 76
    types of, 78
    user education/awareness, 90, 91
    vulnerabilities
        brute force attack, 86, 87
        credential theft, 85
        password weaknesses, 84

INDEX

sensitive information, 84
social engineering
  attacks, 86
AWS, *see* Amazon Web
  Services (AWS)
AWS' Elastic Compute Cloud
  (EC2), 367

B

Backup/restore
  AI/ML integration, 223
  backup (*see* Regular backups)
  blockchain network, 222
  confidential computing, 224
  data-centric security, 223
  homomorphic encryption, 224
  managing practices, 222
  metadata-based security, 223
  methodologies, 203
    differential backups, 204
    full backups, 203
    incremental backups, 203
    mirror backups, 205
    snapshot, 204
  secure backups
    access control, 209
    approaches, 206
    backup types, 211
    compliance/legal
      requirements, 213
    consistent backup, 219
    data deduplication/
      compression, 215

data storage capacity, 208
disaster recovery plans, 211
diversity, 215
documentation/training, 213
emergency recovery
  kits, 218
encryption, 208
end user device, 219
geographical
  advancement, 216
hardware, 220
immutable storage, 214
incremental forever, 218
integrity checks, 210
monitoring/alerts, 212
multiple locations, 207, 208
regular rotation, 217
regular schedule, 206
retention, 220
retention policies, 210
role-based access
  control, 214
secure storage, 212
security, 216
third-party audits, 217
virtual machines, 219
zero trust architecture,
  220, 221
  self-service portals, 224
  unified data protection
    platforms, 225
BGV encryption scheme, 487
Biometric authentication, 82, 83
Blockchain, 481, 490, 491

INDEX

Business/economics, digital extortion
  attack methodology, 334, 335
  disrupted services, 329
  evolution, big business, 330, 331
  financial costs, 340–343
  hackers
    building resilience, 346–348
    Colonial Pipeline attack, 348–351
    increasing costs/ransomware actors, 345, 346
    reducing profitability, 344, 345
  malware, 337–340
  organizational resilience, 352–354
  pricing extortion, 333
  RaaS, 336
  ransomware defense and response, 354–356
  ransomware revenue/profitability, 336, 337
  ransomware value chain, 331, 332

C

California Consumer Privacy Act (CCPA), 9, 16
CASBs, *see* Cloud access security brokers (CASBs)
CCPA, *see* California Consumer Privacy Act (CCPA)
CIA, *see* Confidentiality, integrity, and availability (CIA)
CIA triad, 434
Ciphertext, 47
Cloud access security brokers (CASBs), 369
Cloud computing, 484
  AI/ML, 374
  automation, 374
  case studies, 369, 370, 375
  Cloud adoption, 372, 373
  Cloud-native security, 373
  decentralized security, 375
  Devops, 374
  edge computing, 375
  encryption, 371
  MFA, 370
  network segmentation, 371
  security challenges, 367, 368
Cloud-native security, 373
Cloud security threats, 30
CNSS, *see* Committee on National Security Systems (CNSS)
Committee on National Security Systems (CNSS), 388
Confidential computing
  cloud providers, 365
  data-driven environment, 361
  data privacy/security, 364
  hardware/software, 361
  implementation challenges, 363, 364
  real-world use cases, 362, 363
  security challenges, 365
  threats, 362

INDEX

Confidentiality, integrity, and
    availability (CIA)
  availability, 5
  confidentiality, 4
  encryption, 4
  integrity, 4
  principles, 5
    adhering impact, 7
    artificial intelligence/
      machine learning, 9
    availability, 6
    blockchain, 7
    challenges/consideration, 8
    cloud services/availability, 7
    collaboration/information
      sharing, 10
    confidentiality, 6
    data integrity, 7
    digital transformation, 9
    education/training, 10
    electronic health
      records, 5
    essential framework, 10
    human error, 8
    integrity, 6
    IoT devices, 9
    regulations/data
      protection, 9
    zero trust architecture, 7
  safeguard integrity, 4
COVID-19 pandemic, 307
Cross-site scripting (XSS), 309
Cryptocurrencies, 338
Current threats
  advanced persistent threats
    (APTs), 27
  botnets, 27
  malware, 26, 27
  phishing/social engineering
    attackers, 21
    baiting approach, 24
    emails, 21, 22
    fundamental techniques, 24
    impersonation, 25
    pretexting, 25
    primary technique, 20
    process of, 20
    Quid Pro Quo, 25
    sensitive information, 20
    smishing, 23
    spear phishing, 22, 23
    tailgating, 25
    vishing model, 24
    whaling, 23
  ransomware attack, 26, 27
  steps, 19
  zero-day attacks, 27
Cutting-edge computing, 485
Cyber Kill Chain®, 190, 191
Cybersecurity, 3, 481
  Anthem insurance
    customers, 15
  authentication, 75, 77, 78
    (*see also* Authentication)
  collective responsibility/
    collaboration, 17

INDEX

Cybersecurity (*cont.*)
    data protection
        financial repercussions, 11, 12
        privacy concerns, 11
        theft identification, 12, 13
        vulnerability, 11
    digital security, 3–10
    diverse/profound impacts, 15
    emerging technologies, 16
    Equifax Data Breach, 13
    IRS identity theft, 14
    proactive, 15
    reflection, 15
    strengthening frameworks, 16
    Target data breach, 14
    WannaCry ransomware attack, 14
    Yahoo data breach, 13
Cyberthreat intelligence, *see* Threat intelligence
    concepts, 180
    cyberspace, 177
    definition, 178
    development, 180
    evolution, 179
    intelligence, 176, 177

D

DAC, *see* Discretionary Access Control (DAC)
DarkSide, 348, 412–415, 417
DarkSide malware, 405

Data-driven and evidence-based approach, 394
Data encryption standard (DES), 48
Data loss prevention (DLP), 134, 159
Data privacy and security, 475
DES, *see* Data encryption standard (DES)
Detection techniques
    aberrant occurrence, 155
    administrators/end-users, 144
    AV/EDR system, 146–148
    behavioral analysis, 150
    challenges/solutions, 154–156
    cybersecurity literacy, 143, 144
    dynamic network, 154
    entities, 141
    hallmarks, 144
    implementation/benefits
        anomaly, 149
        comprehensive security, 148–150
        end-users/IT teams, 148
        process anomaly, 148
        process-based monitoring, 150
    interactive training sessions, 145
    investigation
        analysis, 157
        application whitelisting, 159
        DLP solutions, 159
        EDR tools, 158

INDEX

email filtering solutions, 158
files/configurations, 157
honeypot/deception technologies, 159
log analysis, 157
machine learning, 160
MFA implementation, 160
monitoring/report, 158
network traffic patterns, 157, 158
process behavior, 159
sandboxing, 158
security audits/penetration tests, 160
SIEM systems, 160
threat intelligence, 159
legitimate behaviors, 153
limitations, 142
machine learning, 151
malicious software, 146
monitoring, 153
organization, 144
proactive defense, 153
process anomaly
   benefits, 151, 152
   dynamic environments, 150
   network connections, 151, 152
   threats, 150
   unauthorized connections, 152
ransomware, 144
regular updates/refreshers, 145
report suspicious activities, 145
security awareness training, 142, 143
significance vigilance, 156
strategies, 142
swift identification, 153
unexpected file, 154
vital first line of defense, 145
warrants investigation, 154
DevOps, 374
Digital healthcare transformation
analyzing ransomware attacks, 451, 452
case examples/empirical evidence, 470–473
components, 430–432
consumer centric approach, 461
cybersecurity, 429
digital landscape, 474
interim milestones/value management, 462
IR, 453, 454, 456
predictive analytics/ML, 463
preemptive strategies, 463
preventive measures, 439–442
ransomware, 435–438
ransomware detection and response, 443–450
security principles, 434, 435
talent/data challenges, 462
threat intelligence, 456–460
threat landscape, 432–434
transformative security technologies, 465–470

511

Digital security
  AI-enabled multimodal biometrics, 488
  AI/ML, 481
  blockchain technology/security applications, 490, 491
  cloud computing, 484
  edge computing strategies, 483
  encryption
    encrypted machine learning, 487
    homomorphic encryption, 486
    post-quantum, 485
  healthcare, digital transformation, 495, 496
  human-machine adaptation, 489, 490
  IoS, cybersecurity, 491
  quantum computing, data encryption, 485
  remote work environment, 492, 493
  skilled talent, shortage, 493, 494
  ZTA, 482, 483
Discretionary Access Control (DAC), 58
DLP, *see* Data Loss Prevention (DLP)
DKIM, *see* DomainKeys Identified Mail (DKIM)
DMARC, *see* Domain-based Message Authentication, Reporting, and Conformance (DMARC)
Domain-based Message Authentication, Reporting, and Conformance (DMARC), 133, 135
DomainKeys Identified Mail (DKIM), 133
Double extortion, 334

E

ECC, *see* Elliptic curve cryptography (ECC)
E-commerce
  Cyber Kill Chain, 191
  honeypots, 188
  intelligence cycle, 183
  intelligence requirement, 185
  interaction, 188
  malware analysis, 189
  processing/exploitation, 190
  sandboxing, 189
  threat intelligence, 181, 182
E-commerce industry, ransomware attacks
  case studies
    online vendors
      X-cart, 312–315
      Staples, 318–321
      tupperware.com, 309–311
      VF Corporation, 315–318
    cybersecurity, 307, 308
    SQL injections, 309
    XSS, 309
Edge computing, 375, 483

EDR, *see* Endpoint Detection and Response (EDR)
EHRs, *see* Electronic health records (EHRs)
Elastio, 482
Electronic health records (EHRs), 5, 429, 432, 436, 444, 451, 453, 463
Elliptic curve cryptography (ECC), 50
Emerging threats
    AI-driven attacks, 28–30
    cloud computing platforms, 30
    IoT devices, 30
    mitigation strategies, 28
    quantum computing, 31
EML, *see* Encrypted machine learning (EML)
Encrypted machine learning (EML), 487
Encryption, 371
    asymmetric model, 51, 52
    ciphertext, 47
    data protection
        authentication, 52
        confidentiality, 52
        integrity, 52
        nonrepudiation, 53
        values, 54
    decryption, 48
    digital lockdown
        financial ramifications, 115
        inaccessibility, 115
        productivity loss, 115
        reputation, 115
        victim organization, 114
    elements, 70
    email/web security, 132
    symmetric model, 49, 50
    types of, 47
Endpoint Detection and Response (EDR), 69, 125, 147, 158, 168, 169, 246, 464
    adjunct detections, 145
    evolution, 147
    limitation, 147
    recovery process, 240, 241
    threats, 147
Enercon attack, 407
Energy industry
    Amsterdam-Rotterdam-Antwerp refining hubs, 416–418
    Colonial Oil Pipeline
        DarkSide, 413
        digital system, 414
        distribution system, 412
        effectiveness, 415
        human errors, 413
        implementation challenges, 414
    cybercrime, 403
    modern cybersecurity capabilities, 404
    Nordex Group SE/Deutsche Windtechnik
        effectiveness, 408
        implementation challenges, 407, 408

INDEX

Energy industry (*cont.*)
    Windtechnik firms, 407
    wind turbines, 405, 406
    power grids, 403, 404
    Saudi Aramco's Petro Rabigh Complex, 409–412
Enterprise security architecture case studies
    banking sector, 392–394
    components, 387, 388
    definition, 386, 387
    design/implementation, 389–391
    holistic and coherent security, 385
    IT security capabilities, 386
e-skimming attack, 308, 316

F

FAN, *see* Fan Area Network (FAN)
Fan Area Network (FAN), 406
FBI, *see* U.S. Federal Bureau of Investigation (FBI)
FHE, *see* Fully Homomorphic Encryption (FHE)
Firewalls, 371
5G technology, 467
Fully Homomorphic Encryption (FHE), 486, 487

G

GDPR, *see* General Data Protection Regulation (GDPR)

General Data Protection Regulation (GDPR), 9, 16, 444, 481
Google Apps, 367
Granular Backup and Recovery Services (BaaS), 224

H

Health Insurance Portability and Accountability Act (HIPAA), 438, 444
HIPAA, *see* Health Insurance Portability and Accountability Act (HIPAA)
HIPS, *see* Host-based intrusion prevention system (HIPS)
Homomorphic encryption, 487
Host-based intrusion prevention system (HIPS), 244
Houston-based Colonial, 290

I, J

IaaS, *see* Infrastructure as a service (IaaS)
IaC, *see* Infrastructure as code (IaC)
IAM, *see* Identity and access management (IAM)
ICS, *see* Incident Command System (ICS)
ICT, *see* Information and communication technology (ICT)

INDEX

Identity and access management (IAM), 368
IDPS, *see* Intrusion Detection and Prevention Systems (IDPS)
IDS, *see* Intrusion detection system (IDS)
Incident Command System (ICS), 163
Incident response (IR), 163, 453
   analytical phase, 170
   containment/eradication/recovery, 171
   critical stages, 170
   detection/analysis, 170
   log management's role, 164
   multidimensional view, 164
   post-incident analysis, 171
   preparation, 170
   real-world application, 172
   strategic approach, 163
   threat detection, 164
   threats, 163
   tools/technologies, 164–169
Indicator of Compromises (IOCs), 137, 175, 186, 190
Information and communication technology (ICT), 163
Information Technology (I.T.) systems, 290
Infrastructure as a service (IaaS), 367
Infrastructure as code (IaC), 374
Intelligence, 176, 177
Intelligence requirement (IR), 184–186
Internet of Medical Things (IoMT), 432, 433
Internet of Things (IoT), 9, 30, 491, 497
Intrusion Detection and Prevention Systems (IDPS), 164, 166, 243–245
Intrusion detection system (IDS), 131, 137, 171
IOCs, *see* Indicator of Compromises (IOCs)
IoMT, *see* Internet of Medical Things (IoMT)
IoT, *see* Internet of Thing (IoT)

K

Key performance indicators (KPIs), 462, 496
KPIs, *see* Key performance indicators (KPIs)

L

Law enforcement
   collaboration, 273
   collaboration benefits, 271
   logistics of collaboration, 272
   trust/communication, 272
Layered security approach, 289
LockBit ransomware, 471

INDEX

M

MAC, see Mandatory Access Control (MAC)
Machine learning (ML), 9, 16, 151, 160, 223, 374
Malicious insiders, 362
Managed security service provider (MSSP), 296
Mandatory Access Control (MAC), 59, 60
Mazewalker, 404
MFA, see Multifactor authentication (MFA)
ML, see Machine learning (ML)
Modern encryption algorithms, 371
MPC, see Multiparty computation (MPC)
MSSP, see Managed security service provider (MSSP)
Multifactor authentication (MFA), 76, 83, 84, 91, 160, 370
Multimodal biometrics system, 489
Multiparty computation (MPC), 487
Multistep authentication, 79

N

National Institute of Standards and Technology (NIST), 290
NBA, see Network behavior analysis (NBA)
NCA, see UK National Crime Agency (NCA)
Negotiate-down approach, 266
Network-based intrusion prevention system (NIPS), 243
Network behavior analysis (NBA), 244
Network segmentation, 371
NIPS, see Network-based intrusion prevention system (NIPS)
NIST, see National Institute of Standards and Technology (NIST)

O

OSINT, see Open-Source Intelligence (OSINT)
Open-Source Intelligence (OSINT), 106, 188

P

Password-based authentications, 80, 81
Penetration testing, 296, 297
Personal health information (PHI), 451
PGP, see Pretty Good Privacy (PGP)
PHI, see Personal health information (PHI)
Post-quantum digital signature, 485
Pretty Good Privacy (PGP), 134

INDEX

Prevention, ransomware, 119
  comprehensive strategies, 135
    continuous monitoring/
      threat intelligence, 137
    defense-in-depth
      architecture, 136
    education/awareness, 137
    incident response plans, 136
    multi-faceted defense, 135
    regular testing/validation, 138
    risk assessments/
      vulnerability, 136
    risk mitigation, 136
  data backup/disaster recovery
    planning
    business continuity, 126
    components, 126
    disaster recovery plans, 128
    immutable backup
      storage, 127
    incident response
      protocols, 129
    multiple backup copies, 127
    regular backup testing, 128
    regular schedule, 127
    resilience, 126
    write-once, read-many
      (WORM), 127
  email/web security
    anti-phishing capabilities, 133
    awareness training, 134
    content/attachment
      scanning, 133
    data loss prevention, 134
    email spoofing, 135
    filtering solutions, 133
    human element, 133
    implementation, 132
    primary vectors, 132
    URL filtering/web security
      gateways, 134
  employee education/training
    components, 122
    cyber awareness, 122
    cybersecurity awareness, 122
    human error, 122
    regular updates/
      refreshers, 123
    reporting procedures, 123
    role-based training, 123
    security-conscious
      culture, 124
    simulated phishing
      exercises, 123
  network segmentation/access
    controls
    components, 129
    containment/mitigation, 130
    firewalls/access control, 130
    intrusion detection, 131
    isolation/quarantine, 131
    regular security audits/
      assessments, 132
    segmentation/restrict
      access, 131
    segments/zones, 130
  robust endpoint security
    antivirus software, 125

INDEX

Prevention, ransomware (*cont.*)
    application whitelisting, 125
    behavioral analysis, 125
    detection/response, 124
    device encryption, 126
    EDR systems, 125
    firewalls, 125
    implementation, 124–126
    vulnerabilities, 124
  software updates/patch
    management
    automation, 120
    continuous monitoring, 121
    inventory, 121
    prioritization, 121
    procedures, 120
    testing, 121
    unpatched system, 120
    vulnerabilities, 119
Public key encryption
    schemes, 485

Q

Quantum computing, 485
Quantum computing threats, 31
Quantum-resistant encryption
    schemes, 485

R

RaaS, *see* Ransomware-as-a-
    service (RaaS)
Ransomware, 119, 288

archives attack, 101
backup/restore (*see* Backup/
    restore)
communication/extortion
  assessing
    communication, 117
  collaboration/
    transparency, 118
  navigation, 117
  negotiation, 116
  phases, 116
  pressure tactics, 116
  seeking external
    assistance, 117
  strategic response, 117
cryptolocker, 102
DarkSide, 103
definition, 97
deployment
  critical purpose, 113
  decryption key, 114
  encryption process, 112
  legal/regulatory
    implications, 114
  payment instructions, 113
  ransom demand, 113
  risky decision, 114
  scanning algorithms, 112
  threat, 113
  unleashing encryption, 112
detection (*see* Detection
    techniques)
disruptive/lucrative attacks, 103
drawbacks

INDEX

cyber insurance, 268
data criticality, 267, 274
data recovery, 267
legal counsel, 269, 270
operational disruption, 268
public relations/brand equity, 269
ransom payment feasibility, 267
third-party stakeholder's data, 268
encrypting target files, 114–116
evolution, 97, 98
GPCode attack, 101
individuals/organizations, 99, 100
initial access
   drive-by downloads, 108
   exploit vulnerabilities, 108
   malvertising, 108
   methods, 107
   phishing emails, 107
   RDP exploits, 109
   social engineering, 109
   spear phishing, 108
   supply chain attack, 109
   unpatched software, 108, 109
   watering hole attacks, 108
investigation process, 275
lateral movement/privilege escalation
   active directory (AD), 110
   evading detection, 111
   lateral movement, 111, 112
   pass-the-hash attacks, 110
   remote access tools, 110
   system privileges, 111
   techniques, 110
law enforcement, 271
legal implications, 260–262
lifecycle, 105
locky/cryptowall, 102
mitigation strategies, 99
moral considerations, 262–264
national security/economic stability, 103
NotPetya, 102
reconnaissance/target selection
   advanced persistent threat (APTs), 106
   attractive targets, 106
   data breaches, 106
   geopolitical objectives, 107
   OSINT techniques, 106
   perceive inability, 107
   public image, 107
   social engineering, 105
   target rich environment, 106
   techniques, 105
   valuable data, 106
recovery (*see* Recovery process)
risks/benefits, 264–266
unlawful activity and violate laws, 260
wannacry, 102
See also Prevention, ransomware
Ransomware-as-a-service (RaaS), 98, 103, 310, 331, 336, 415

519

INDEX

RBAC, *see* Role-based access control (RBAC)
RDP, *see* Remote Desktop Protocol (RDP)
Recovery planning flowchart, 230
Recovery process
    access controls, 240
    attack verification/reaction, 231, 232
    backup activities, 236, 237
    characteristics, 234
    cyber-security event, 230
    damage assessment, 232
    decryption activities, 236
    delivery/host activity, 241
    divided networks, 239
    EDR solutions, 240, 241
    hardening controls, 237
    honeypot system, 247, 248
    IDS/IPS systems, 242–244
    incident response playbooks, 249, 250
    indicators, 230
    isolation actions, 231, 232
    key leaders/decision-makers, 233
    microperimeters, 240
    network segmentation, 239, 240
    options, 234–236
    overview of, 230
    production network, 231
    ransom payment, 249
    report details, 233
    review/improve security, 238
    segmentation, 238
    SIEM solutions, 244, 245
    significant threats, 229
    Splunk/QRadar/LogRhythm, 245
    strategies/tools, 250
    UBA tools, 246
    vulnerabilities, 238
    zero-trust approach, 239
Regular backups
    benefits, 199
    business continuity, 201
    compliance, 201
    data backups, 202
    data restoration, 199
    integrity, 200
    mitigating ransom demands, 200
    security and strategy, 201
Remote Desktop Protocol (RDP), 109
Rivest–Shamir–Adleman (RSA), 50
Role-based access control (RBAC), 60, 61, 130
RSA, *see* Rivest–Shamir–Adleman (RSA)
Russian-based cryptocurrency, 261
Ryuk, 405

S

SaaS, *see* Software-as-a-service (SaaS)
SABSA, *see* Sherwood Applied Business Security Architecture (SABSA)

INDEX

Salesforce, 367
SCADA ethernet network, 406
SEC, *see* States Securities and Exchange Commission (SEC)
Security
  cyberattacks
    breached sector, 284
    data breaches, 283, 285
    disclosed incidents, 285
    incidents, 286
    ransomware, 288, 289
    social engineering, 288
    types of attacks, 287
  implementing enhancements
    cybersecurity-centric culture, 292, 293
    cybersecurity framework, 290, 291
    monitoring and management threats, 295
    penetration testing/vulnerability scanning, 296, 297
    policy adjustments, 290
    technological upgrades, 293–295
Security Information and Event Management (SIEM), 67, 137, 160, 166, 167, 244, 245, 295
Security measures
  antivirus software, 66
  applications/approaches, 69
  approaches, 71
  data loss prevention, 67
  different models, 64
  digital signatures, 65
  endpoints, 69
  firewalls, 65
  hashing process, 65
  integrity checks, 64
  intrusion detection/prevention systems, 66
  network segmentation, 69
  patch management, 68
  regular security audits, 68
  SIEM tools, 67
  vulnerabilities, 68
Security orchestration, automation, and response (SOAR) systems, 368
Sender Policy Framework (SPF), 133
Service-level agreements (SLAs), 224
Service Organization Control (SOC), 290
Sherwood Applied Business Security Architecture (SABSA), 388
SIEM, *see* Security Information and Event Management (SIEM)
SIEMaaS, *see* SIEM-as-a-Service (SIEMaaS)
SIEM-as-a-Service (SIEMaaS), 295
SLAs, *see* Service-level agreements (SLAs)

INDEX

SOAR systems, *see* Security orchestration, automation, and response (SOAR) systems
SOC, *see* Service Organization Control (SOC)
Social engineering, 288
Software-as-a-service (SaaS), 367
Software-defined networking (SDN), 371
SPF, *see* Sender Policy Framework (SPF)
States Securities and Exchange Commission (SEC), 316
STIX, *see* Structured Threat Information Expression (STIX)
Structured Threat Information Expression (STIX), 183
Stuxnet, 405, 409
Supply chain attacks, 289
Symmetric encryption model, 49, 50

T

TEEs, *see* Trusted execution environments (TEEs)
TLS, *see* Transport layer security (TLS)
The Open Group Architecture Framework (TOGAF), 388
Threat detection, 164, 172
Threat intelligence, 175, 456
    command and control (C2), 186
    data science, 178
    definition, 178
    e-commerce, 181–183, 185, 186
        Cyber Kill Chain®, 190, 191
        honeypots, 188
        interaction, 188
        IR information, 185, 186
        malware analysis, 189
        processing/exploitation, 190
        sandboxing, 189
    evolution
        history, 178
        industrial revolution, 179
        post-war intelligence, 179
        renaissance, 179
        world war, 179
    Intelligence cycle, 183, 184
        analysis/production, 184
        dissemination and integration, 185
        evaluation and feedback, 185
        planning/targeting, 184
        preparation and collection, 184
        processing/exploitation, 184
    malware, 186–188
    operational intelligence, 182
    OSINT data, 188
    risk management processes, 180
    strategic intelligence, 182
    tactical intelligence, 182, 183
Threat intelligence platforms (TIPs), 457

INDEX

Threats
  current threats, 19–27
  cybercrime, 33, 34
    critical distinctions, 40–42
    emerging threats, 28–31
    law enforcement/judicial
      process, 41
    malware campaigns, 35–37
  mitigation strategies, 31
    adversarial techniques, 32
    AI-Powered Defense
      Systems, 31
    collaboration/information
      sharing, 33
    continual monitoring/
      adaptation, 33
    data security/privacy
      measures, 32
    education/awareness
      approaches, 32
    regular assessment/
      penetration testing, 32
    regulatory compliance, 33
    vital strategic activities, 31
  motivations, 41
  state-sponsored attacks
    critical distinctions, 40–42
    cyber deterrence/warfare, 39
    cyber operations, 36
    espionage, 38
    government secrets, 38
    industrial espionage, 38
    motivations, 38, 40
    objectives, 36
    political/strategic
      motivation, 38
    process of, 37
    sabotage/disruption, 39
    zero-day approach, 37
TIPs, *see* Threat intelligence
  platforms (TIPs)
TOGAF, *see* The Open Group
  Architecture
  Framework (TOGAF)
Transport layer security
  (TLS), 49, 134
Triple DES (Data Encryption
  Standard), 486
Triton, 405
Trusted execution environments
  (TEEs), 224, 361, 363
Two-factor authentication,
  79, 81, 87

U

UBA, *see* User Behavior
  Analytics (UBA)
UEBA, *see* User and Entity Behavior
  Analytics (UEBA)
UK National Crime Agency
  (NCA), 271
University of Vermont (UVM)
  Health Network, 434
User and Entity Behavior Analytics
  (UEBA), 158

523

User Behavior Analytics (UBA), 246
U.S. Federal Bureau of Investigation (FBI), 309
UVM Health Network's systems, 442

V

Virtual local area networks (VLANs), 239
Virtual private networks (VPNs), 49, 371
Virtual reality (VR), 468
VLANs, *see* Virtual local area networks (VLANs)
VPNs, *see* Virtual private networks (VPNs)
VR, *see* Virtual reality (VR)

W

WIPS, *see* Wireless intrusion prevention system (WIPS)
Wireless intrusion prevention system (WIPS), 243

X, Y

X-Cart ransomware attack, 314
XSS, *see* Cross-site scripting (XSS)

Z

Zero trust architecture (ZTA), 482
Zero trust model, 464
Zero trust network access (ZTNA), 482
ZTA, *see* Zero trust architecture (ZTA)
ZTNA, *see* Zero trust network access (ZTNA)

Printed in the United States
by Baker & Taylor Publisher Services